SPRINGER TRACTS IN MODERN PHYSICS

Ergebnisse
der exakten Natur-
wissenschaften

Volume **68**

Editor: G. Höhler

Associate Editor: E. A. Niekisch

Editorial Board: S. Flügge J. Hamilton F. Hund
H. Lehmann G. Leibfried W. Paul

Springer-Verlag Berlin Heidelberg GmbH 1973

Manuscripts for publication should be addressed to:

G. HÖHLER, Institut für Theoretische Kernphysik der Universität, 75 Karlsruhe 1, Postfach 6380

Proofs and all correspondence concerning papers in the process of publication should be addressed to:

E. A. NIEKISCH, Kernforschungsanlage Jülich, Institut für Technische Physik, 517 Jülich, Postfach 365

ISBN 978-3-662-15566-0   ISBN 978-3-540-46942-1 (eBook)

DOI 10.1007/978-3-540-46942-1

Solid-State Physics

Contents

Nuclear Magnetic Double Resonance — Principles and Applications in Solid-State Physics

Dankward Schmid

Contents

I. Introduction

During the last two decades nuclear magnetic resonance (NMR) has become an increasingly important experimental technique for obtaining various information in physics and chemistry. In the first place, it has become possible to determine nuclear magnetic dipole moments and electric quadrupole moments with high accuracy. Second, the local magnetic fields and electric field gradients, and thus for instance the structure of molecules can be investigated, if the moments of the involved nuclei are known. Studies of relaxation phenomena yield information about internal interactions and motions of the nuclei.

However, there are severe limitations of the sensitivity of conventional NMR experiments. To obtain a detectable signal at room temperature and at a magnetic field of 10 kG at least 10^{18} protons are required, even at optimum conditions. Even more nuclei are required, if they have a smaller gyromagnetic ratio or if the NMR line is broadened by dipolar interaction in solids. Since there are numerous possible applications of great practical importance, there have been intense efforts to increase the sensitivity of NMR. The methods to improve the sensitivity can be divided into the following different groups:

a) Enhancement of the Nuclear Polarization

Since the signal strength is proportional to the static magnetization, which is given by Curie's law, the sensitivity increases if the static magnetic field, B_0, is chosen as high as possible and the temperature as low as possible. The increase of B_0 is limited by the magnets available, yet the use of high-field superconducting magnets has yielded a considerable improvement of the sensitivity. Lowering the temperature, on the other hand, often causes unfavorable long relaxation times. The radio frequency (rf) field then destroys the thermal equilibrium between the spin system and the lattice, and consequently the net magnetization decreases.

The polarization of the nuclei to be observed can also be enhanced, if they are coupled to another spin species (electrons or nuclei with a larger γ) and if one saturates the NMR transition of this second species by irradiation of a strong resonating rf field. This is known as dynamic nuclear polarization (Overhauser effect [1, 2], solid state effect [3, 4]). Dynamic nuclear polarization can also be achieved by irradiation of polarized light (optical pumping [5, 6]).

b) Indirect Detection of NMR via Quantum Transformation

A weak NMR can often be observed because it changes another more intense signal. The ENDOR method uses this possibility [7, 8]. Here the NMR is identified because of its influence to the electron spin resonance signal of a paramagnetic impurity, which interacts with the nucleus. The optical double resonance technique [9, 10, 11] represents another example. In both cases the enhancement of the sensitivity results mainly from a "quantum transformation", that is, an NMR transition is observed by looking at the absorption or emission of quanta of much higher energy.

The methods *a* and *b* have been reviewed extensively [2, 3, 8, 11]. In this paper a third method will be discussed:

c) Detection of NMR by Means of Energy Exchange between Spin Systems (Nuclear Double Resonance)

In many diamagnetic compounds the interesting nuclear spin species have too low concentrations or too small magnetic moments to be detected by conventional NMR. But many compounds contain in addition another nuclear spin species with a high concentration and large magnetic moments (for instance ^{1}H, ^{19}F, ^{23}Na, ^{31}P), which yield large NMR signals. The *nuclear double resonance method* (DNMR) uses this abundant spin species to observe the rare spin NMR indirectly [12, 13, 14]. The basic idea of this method is as follows: First one places the ensemble of abundant spins I[1] in a highly ordered state[2], in which

---

[1] Throughout this paper the abundant spin species will be referred to as *I* spins, the rare species as *S* spins.

[2] A highly ordered state of a spin system usually means that there exists a large polarization or magnetization along the direction of a given magnetic field. We are speaking here about order instead of magnetization since the latter does not characterize the state of the system unambiguously. A spin system in zero magnetic field for instance does not possess a magnetization at thermal equilibrium. Likewise, a macroscopic magnetization does not exist in a spin system which has been demagnetized adiabatically from thermal equilibrium in a high magnetic field. Yet both states are fundamentally different: If we apply a magnetic field adiabatically to the first system the magnetization will build up only in times of the order of the spin lattice relaxation time. In the second case, the magnetization will increase immediately as the magnetic field increases and will finally regain its full initial size, if relaxation is negligible during the demagnetization. The entropy of the system is not changed by adiabatic variations of the magnetic field. The system has the same degree of order in the adiabatically demagnetized state as before. Throughout this paper we will therefore frequently use the concept of order of a spin system instead of magnetization to characterize its state.

it remains for a time of the order of the spin lattice relaxation time. Subsequently the rare spin system S is brought into a state of maximum disorder by irradiation of an rf field near the resonance frequency. If it is possible to couple the two spin systems by a suitable choice of the experimental parameters, order within the I spin system can be destroyed by irradiation of an rf field near the S spin resonance frequency and thus the S spin resonance can be detected.

There are two major points of interest in DNMR research: First, the coupling between the spin systems can be investigated by studying the sensitivity and dynamics of the DNMR. One obtains information about the behavior of spin systems in strong rf fields [15], which in turn is useful e.g. for the understanding of relaxation phenomena in solids or of the dynamic nuclear polarization. In the work of the groups around Hahn and Slichter [12, 13, 16 – 19] these problems have been studied extensively. Second, during the last years this method has been also applied successfully to investigate the structure of point defects in solids [14, 20 – 28]. Furthermore DNMR has yielded interesting results in the analysis of phase transitions, for instance in KH_2PO_4 [29, 30]. Thus there is a wide field for application of DNMR in solid state physics.

This paper is divided into the following chapters: Part two describes the basic ideas of the DNMR experiment in an extremely simplified way. This qualitative description intends to make the important steps of the DNMR experiment as evident as possible. In part III various experimental methods are discussed. In part IV results are presented, which have been obtained by application of DNMR in the field of solid state physics. Part V contains a short review of the theory of DNMR, especially its dynamics. However, we do not intend to give a complete description, but rather want to indicate how the principal features of the DNMR can be explained by use of the thermodynamics of spin systems. In part VI the influence of spin diffusion is considered. Finally in part VII the possible applications and the information gained by use of DNMR are summarized briefly.

II. Simplified Description of the Nuclear Double Resonance Method

II.1. A Simple Thermodynamic Analogon

In order to explain the principles of DNMR we want to discuss a simple thermodynamic model experiment [13] (see Fig. 1). Consider two bodies which are connected by a heat conducting rod. One of them, having a small heat capacity, represents the S spin system, which absorbs the

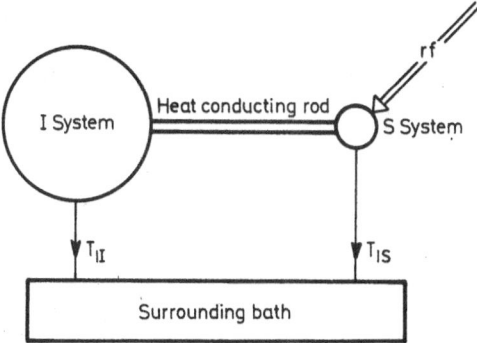

Fig. 1. Thermodynamic model illustrating the DNMR experiment.

NMR rf field only weakly, the other one of large heat capacity represents the abundant I spins. The S spin system can be heated externally by irradiation of a resonating rf field. However, the heat capacity of the S system is too small, to allow a direct measurement of the increase of its temperature. But if the S system is heated long enough and the thermal conductivity is sufficient, the temperature of the I system will increase gradually, provided that it is only weakly coupled to the surrounding bath (lattice). Lurie and Slichter have pointed out that the experiment can be performed in different ways. Either the S system can be heated continuously. It then will reach a constant high temperature and the rate of increase of the I system temperature depends on the thermal conductivity of the rod and the heat capacity of the I system, but not on the heat capacity of the S system. Or one can break the thermal contact between the two systems, heat the S system to a known high temperature and connect them again. After a sufficiently long time the I system, the S system, and the rod reach a common temperature, which depends only on their respective heat capacities and initial temperatures. Since the heat capacity of the S system is small, the temperature of the I system increases only by a small amount. However, one can repeat the cycle many times and thus get a measurable increase of the I system temperature.

Using this model experiment one can illustrate all the quantities which can be measured in a DNMR experiment. First, an increase of the I spin temperature indicates that the S system has been heated, which in turn means that NMR absorption has taken place in the S spin system. The enhancement of the sensitivity results from the fact that the I system integrates over many S spin resonance absorption cycles. Second, one can also measure the ratio of the heat capacities of the two systems and the thermal conductivity of the rod, i.e. the thermal contact between the systems. The latter quantities can be varied experimentally over a wide range and can be calculated by use of the thermodynamics of spin systems. DNMR therefore gives a possibility to compare this theoretical concept with the experiment. This point will be discussed in part V.

II.2. The Rotating Frame

To describe the behaviour of spins in a magnetic field it is very convenient to use a rotating coordinate system [31]. Since it is indispensable for the description of DNMR, its most important features shall be discussed in this section. This discussion will be restricted to classical aspects. The quantum mechanical treatment will be necessary only for the theory of DNMR and is therefore put back to part V.

Consider a free nucleus with a spin $I\hbar$ and a magnetic moment $\mu = \gamma\hbar I$ (γ is the gyromagnetic ratio) in a magnetic field B. The magnetic field produces a torque on the nucleus of the amount $D = [\mu \times B]$. The equation of motion is therefore

$$d\mu/dt = \gamma[\mu \times B] . \tag{II.1}$$

The solution of Eq. (II.1) is most readily obtained by transforming it into a frame of reference which rotates with the angular frequency ω relative to the laboratory frame around the common z-axis. Consider an arbitrary vector function F. According to the rules of vector algebra the time derivatives with respect to the laboratory and the rotating frame, dF/dt and $(\partial F/\partial t)_{rot}$ are related to each other by the equation

$$dF/dt = (\partial F/\partial t)_{rot} - [F \times \omega] . \tag{II.2}$$

Eq. (II.1) therefore transforms to the rotating frame as

$$(\partial\mu/\partial t)_{rot} = \mu \times (\gamma B + \omega) . \tag{II.3}$$

The motion of the magnetic moment in the rotating frame obeys therefore the same rules as in the laboratory frame, if the magnetic field B is replaced by an effective field, $B_{eff} = B + \omega/\gamma$.

If $B = B_0$ is time independent we can solve Eq. (II.3) immediately by choosing $\omega = -\gamma B_0$. Since the time derivative in this reference frame vanishes, μ remains fixed with respect to its axes. Relative to the laboratory coordinate system it therefore precesses at an angular frequency $\omega_0 = -\gamma B_0$ around the direction of B_0.

There is a simple intuitive explanation for the effective field: It is well known that μ precesses in a static magnetic field B with an angular frequency $\omega_0 = -\gamma B$. With respect to a coordinate system which rotates at exactly this frequency around B, the magnetic moment therefore is fixed, i.e. it behaves as if no magnetic field was acting on it. If the rotating frame rotates at a frequency $\omega \neq \omega_0$, μ precesses in this frame at a frequency $(\omega_0 - \omega)$, that is as if it would be exposed to an effective field $B_{\text{eff}} = B + \omega/\gamma$.

The advantage of a rotating frame shows up most strikingly, if the influence of alternating magnetic fields is considered. In most magnetic resonance experiments a linearly polarized alternating magnetic field, for instance $B_x(t) = 2B_1 x \cos \omega t$ is applied perpendicular to the static magnetic field $B_0 z$. (x, y and z are the unit vectors corresponding to the axes of the laboratory frame.) Such a linearly polarized field can be divided into two rotating components, one of them rotating in the sense of the Larmor precession the other opposite to it. If ω is close to the resonance frequency ω_0, and if ω_0 is large, the counter-rotating component may be neglected in most cases. Then the total magnetic field is given by the sum of a static field, $B_0 \cdot z$, and a rotating field, $B_1(t) = B_1(x \cos \omega t - y \sin \omega t)$. The time dependence of $B_1(t)$ can be eliminated by transforming it into a rotating frame (x', y', z'), which rotates around the z-axis at an angular frequency, $-\omega \cdot z$, and whose x'-axis coincides with B_1. The equation of motion in this coordinate system becomes

$$(\partial \mu / \partial t)_{\text{rot}} = \mu \times ((B_0 - \omega/\gamma) z' + B_1 x'), \tag{II.4}$$

(x', y' and z' are the unit vectors corresponding to the axes of the rotating frame.) In this frame μ precesses around the static effective field

$$B_{\text{eff}} = (B_0 - \omega/\gamma) z' + B_1 x', \tag{II.5}$$

(see Fig. 2).

The problem is particularly simple if ω is exactly at resonance, $\omega = -\gamma B_0$. The effective field then simply is $B_{\text{eff}} = B_1 \cdot x'$. Suppose, before switching on $B_1(t)$ the magnetic moment was aligned along the z-axis. If one now turns on B_1 the magnetic moment precesses in the rotating frame in a plane perpendicular to B_1 pointing periodically parallel or antiparallel to B_0. If B_1 is switched on only for a short time τ, defined by the relation $\gamma B_1 \tau = \pi/2$, then the magnetic moment coincides with the y'-axis of the rotating frame (see Fig. 3). This is known as a 90° pulse. Subsequently the vector of the magnetization is fixed in the rotating frame, that is in the laboratory frame it precesses at an angular frequency $\omega_0 = -\gamma B_0$ in a plane perpendicular to B_0. In a pick up coil, whose axis is perpendicular to B_0, it therefore induces an rf voltage at the frequency ω_0. This voltage should exist indefinitely, if there were no other perturbations. However, it is well known that the transverse magnetization, M_t decays rapidly (free induction decay) and

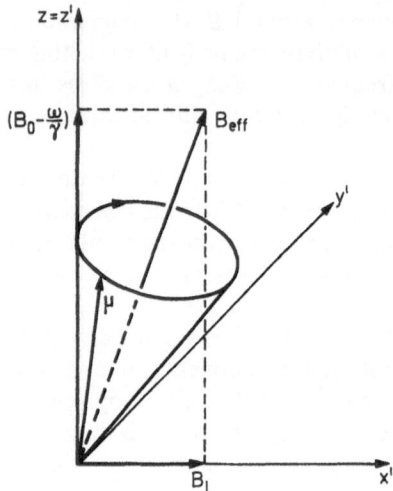

Fig. 2. Cone of precession of a magnetic moment μ in the rotating frame.

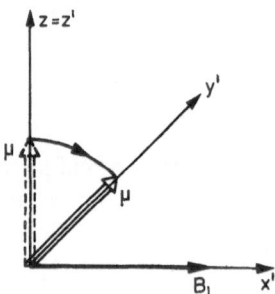

Fig. 3. 90° pulse. The magnetic moment precesses in the $y'z$-plane of the rotating frame around the effective field B_1. B_1 is switched off, when μ coincides with the y'-axis.

that the longitudinal magnetization, M_l parallel to $\boldsymbol{B}_0$ increases again to its thermal equilibrium value, M_0. In order to give a phenomenological description of the approach to the equilibrium situation ($M_t = 0$, $M_l = M_0$) Bloch introduced two time constants, T_2 and T_1 respectively, and modified the equations of motion in the rotating frame in the following way:

$$\mathrm{d}M_{z'}/\mathrm{d}t = (M_0 - M_{z'})/T_1 + \gamma[\boldsymbol{M} \times \boldsymbol{B}_{\mathrm{eff}}]_{z'},$$
$$\mathrm{d}M_{x'}/\mathrm{d}t = -M_{x'}/T_2 + \gamma[\boldsymbol{M} \times \boldsymbol{B}_{\mathrm{eff}}]_{x'}, \qquad\qquad (\mathrm{II}.6)$$
$$\mathrm{d}M_{y'}/\mathrm{d}t = -M_{y'}/T_2 + \gamma[\boldsymbol{M} \times \boldsymbol{B}_{\mathrm{eff}}]_{y'}.$$

The substitution of $\boldsymbol{M}$ instead of $\boldsymbol{\mu}$ accounts for the fact that one does not observe the magnetic moment of single spins but always an ensemble average of a large number of spins. The T_1-process describes the approach of the population of the Zeeman-levels to its equilibrium state, a Boltzmann distribution, which is characterized by the lattice temperature. This requires an energy exchange between the spin system and the lattice. The transverse magnetization on the other hand results from the fact that in a classical picture the spins precess preferentially with a definite phase. A statistical phase distribution can be reached because of interaction of the spins with each other. Therefore the decay of the transverse magnetization, the T_2-process, does not require an energy exchange with the lattice.

II.3. Spin Locking

Bloch's Eqs. (II.6) give a very good explanation of NMR experiments in liquids and gases. In NMR experiments in solids with strong rf fields Redfield [15] discovered in 1955 severe discrepancies between the experimental results and Bloch's theory: In the absence of resonating rf fields the transverse magnetization in solids decays usually in times of the order of 100 μsec. Although this decay is generally non-exponential [33] it is often also characterized by a time constant T_2. T_2 measures the rate of the loss of phase coherency between the spins because of their mutual dipole interaction. It is caused by the spatial fluctuations of the magnetic field in the sample and can be defined as

$$T_2 \approx 1/\gamma B_{\text{loc}} . \tag{II.7}$$

B_{loc} is approximately the root mean square magnetic field at a site of one nucleus due to the dipolar fields of all his neighboring spins. A precise definition will be given in part V.

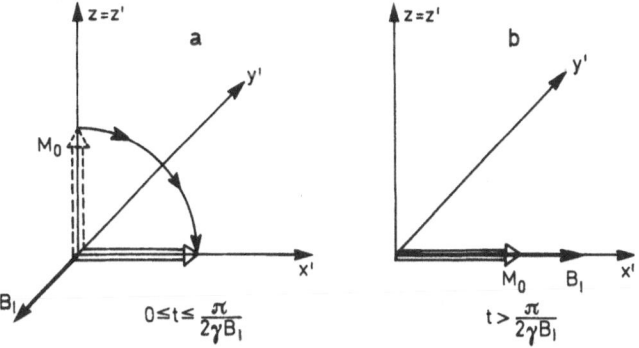

Fig. 4. Spin locking. a The magnetization is turned from the z-direction to the x'-direction, using a 90° pulse. b Following a sudden 90° phase shift of B_1, M_0 is parallel to B_1 in the rotating frame

Without rf fields the transverse magnetization decays rapidly compared to the spin lattice relaxation, which typically has time constants
between 1 sec and 1000 sec. However, it is possible to tilt the magnetization into the xy-plane in such a way that it precesses synchronous
and in phase with the rotating rf field B_1. This can be accomplished, for
instance, if immediately after a 90° pulse B_1 is not switched off, but if
instead its phase is shifted suddenly by 90° so that B_1 is lined up along
the magnetization vector. If B_1 is large enough ($B_1 \gg B_{loc}$) one observes
that the transverse magnetization decays only slowly, in times of the
order of T_1 (see Fig. 4). This phenomenon is called spin locking and it
is one of the most important facts which make the DNMR experiment
feasible.

II.4. Rotary Saturation

The Hamiltonian of the spin system in the rotating frame is time independent and equivalent to the Hamiltonian of a spin system in a static
magnetic field in the laboratory frame. (The nonsecular parts of the dipolar interaction can be neglected in most cases. This will be discussed
in part V.) B_{eff} replaces the static magnetic field, and its direction defines
the axis of quantization. Exactly at resonance the x'-axis of the rotating
frame therefore is equivalent to the laboratory frame z-axis, and B_1
becomes the static field equivalent to B_0 in the laboratory frame. If
B_0 is modulated at a frequency $\omega_a = \gamma B_1$, it induces resonance transitions
in the rotating frame (see Fig. 5). This was demonstrated by Redfield [15]
in an experiment which he called rotary saturation. Franz and Slichter
[34] extended his work and proved that the experimental results of rotary
saturation can be explained in detail by analogy with NMR experiments

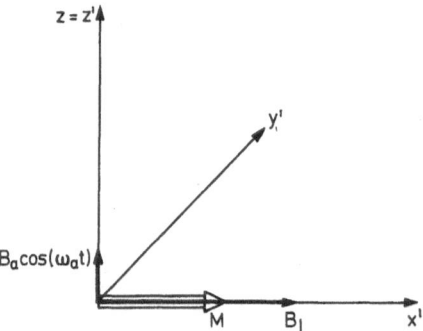

Fig. 5. Rotary saturation. The transverse magnetization M of a spin locked system is
destroyed, if an audio field B_a is applied at the resonance frequency, $\omega_a = \gamma B_1$

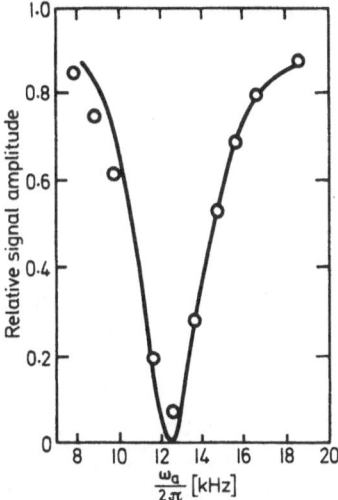

Fig. 6. Rotary saturation in CaF$_2$ [34]. The ^{19}F nuclei were spin locked in an rf field, $B_1 = 3.05 \pm 0.10$ G. The fraction of the remaining magnetization is plotted versus the frequency $\omega_a/2\pi$ of the audio field, $B_a = 0.08 \pm 0.005$ G

in the laboratory frame. Fig. 6 shows an example from their work. The ^{19}F nuclei in an CaF$_2$ crystal were spin locked with an rf field $B_1 = 3.1$ G. Subsequently an audio frequency field B_a was applied parallel to the $z = z'$ direction. ω_a was varied and in Fig. 6 the remaining fraction of the magnetization after B_a was turned off is plotted as a function of ω_a. It shows a strong destruction of the magnetization at the resonance frequency $\omega_a/2\pi = \gamma B_1/2\pi = 12.4$ kHz.

II.5. Nuclear Double Resonance (DNMR)

The rotary saturation experiment leads to the basic idea of the DNMR. Instead of applying B_a externally, it is possible to create it internally by use of a second spin species in the sample. Suppose, there are two spin species in a sample (e.g. ^{6}Li and ^{7}Li in lithium metal). Then it is possible to spin lock the abundant I spins (^{7}Li) with an rf field B_{1I}, without disturbing the rare spin species $S(^6$Li) since their resonance frequencies are different. Now one applies a second rf field B_{1S}, whose frequency is at resonance with the S spins, $\omega_S = \gamma_S B_0$. Then the S spins precess in a frame of reference, which rotates at the frequency ω_S with respect to the laboratory frame, with the angular frequency $\omega_{1S} = \gamma B_{1S}$ around B_{1S}. Thus, they create magnetic fields in the sample, whose z-components oscillate at the frequency ω_{1S}. If one adjusts the rf amplitudes properly to match the condition

$$\gamma_S B_{1S} = \gamma_I B_{1I}, \tag{II.8}$$

the situation is the same as in the case of rotary saturation. The oscillating S spin dipole field destroys order within the spin locked I spin system. NMR transitions within the S spin system are indicated by a loss of order of the I spin system. Eq. (II.8) is often referred to as Hahn's double resonance condition.

The DNMR process is different from rotary saturation in the following regard: The decay of order of the I spin system is accompanied by an increase of order in the S spin system. After a certain exchange time both of them are in a partially spin locked equilibrium state. (This exchange time is of the order of T_2, if the double resonance condition is matched. In most cases T_2 is of the same order of magnitude for the I and S spins, and in solids it is always small compared to T_1. For the present qualitative discussion it is only important that the exchange time is short compared to T_1. This will be discussed in more detail in part V.) The exchange of order is often called a temperature exchange. If both spin species are present at approximately the same concentration, a single exchange process causes a measurable loss of order of the initially spin locked system. However, if the S spins are very rare the loss of order of the I spin system is hardly detectable. But, since the lifetime of the metastable state is of the order of T_1, which is much longer than the time constant for the exchange process, the exchange of order can be repeated many times. If the equilibrium between both spin systems is reached, B_{1S} can be switched off suddenly. Then the order within the S spin system decays within a time of the order of T_2. Subsequently the S spin system can be heated again by irradiation of another B_{1S} pulse. In this way the double resonance process can be repeated many times, and the sensitivity of the detection of an S spin NMR increases according to the number of repetitions.

In conclusion of this section a very simple comprehensive description of the DNMR process shall be given, using two spin species, $I = 1/2$ and $S = 1/2$ as an example. If $\gamma_S \neq \gamma_I$ mutual spin flip flop processes are not allowed in the laboratory frame without interaction with the lattice, since the energy is not conserved. In a DNMR experiment there is exchange of order but energy is conserved in a first order approximation. To make this point clear one must define the concept of order more precisely and state the difference between the spin locked (ordered) I spin system and the disordered S spin system. In either case the Zeeman levels in the laboratory frame, $|m_I\rangle = |\pm \tfrac{1}{2}\rangle$ and $|m_S\rangle = |\pm \tfrac{1}{2}\rangle$, are equally populated. Thus the magnetization in the z-direction vanishes. However, the S spins precess around B_0 with statistical phases, whereas the spin locking B_{1I} field creates a coherence in the precession of the I spins. It is this coherence which stores the order of the I spin system. The exchange of order is accomplished by mutual phase changes. The z-components of the magnetization M_{Iz} and M_{Sz} are not changed, thus in a first order approximation no energy is exchanged. In the rotating frame however, these phase changes correspond to mutual flip flop processes with respect to the effective fields B_{1I} and B_{1S}, respectively, which is often referred to as cross relaxation in the rotating frame. If the double resonance condition Eq. (II.8) is satisfied, the energy exchange between the two systems is possible in the rotating frame, and the total energy is conserved.

II.6. Spin Temperature in the Rotating Frame

The following section gives a comparison between the model experiment and the individual steps of the DNMR experiment. The I spin system in the spin locked state can be considered as a thermal reservoir because of the following reasons: Immediately after the spin locking process the energy levels in the rotating frame are populated according to a Boltzmann distribution, characterized by a spin temperature in the rotating frame, θ'. It measures the degree of order of the spin system. In the course of the spin locking process the order of the spin system is conserved. Thus the Boltzmann distribution in the rotating frame is the same as the initial distribution in the laboratory frame. The spin temperature of the spin locked system is much lower than the equilibrium temperature in the laboratory frame θ_l (lattice temperature), since the splitting of the Zeeman levels in the rotating frame, $\Delta E' = \gamma \hbar B_1$, is much smaller than the splitting in the laboratory frame, $\Delta E = \gamma \hbar B_0$. In solids the dipolar interaction couples each spin strongly to its neighbors. Such a system of many interacting particles cannot be treated exactly. But, following Redfield, it is reasonable to assume that the most probable state of the system is a canonical distribution in the rotating frame. For such a canonical system the rules of conventional thermodynamics are valid. Thus they can also be applied in the rotating frame. This procedure yields a qualitative explanation of the spin locking for nuclei with $I = \frac{1}{2}$.

Suppose that at the beginning of the spin locking the full equilibrium magnetization, $M_0 = \gamma \hbar n_0 / 2$, was lined up along B_1 (n_0 is the population difference of the Zeeman levels at thermal equilibrium in the laboratory frame). The energy of the system in this initial state is

$$E_i = -\tfrac{1}{2} \gamma \hbar n_0 B_1 . \tag{II.9}$$

In order to destroy the transverse magnetization this energy must be supplied to the spin system. In the limit of zero spin lattice relaxation this energy can come only from the internal dipole coupling energy, i.e. roughly speaking the magnetic moments of the nuclei must be aligned preferentially to their respective local field. Hereby the energy

$$E' = \tfrac{1}{2} n' \gamma \hbar B_{loc} \tag{II.10}$$

becomes available. n' is the population difference of states with spins parallel or antiparallel to the local field. The energy of the final state is then given by

$$E_f = -\tfrac{1}{2} \gamma \hbar (n_f B_1 + n' B_{loc}), \tag{II.11}$$

where n_f is the population difference of the Zeeman levels in the final state.

Similarly the entropy of the system can be calculated. If N is the total number of spins and k the Boltzmann constant, the values of the initial and final entropy are [35]

$$S_i = -\tfrac{k}{2} \left[\ln(\pi N/2) + n_0^2/N \right], \tag{II.12}$$

and

$$S_f = -\tfrac{k}{2} \left[\ln(\pi N/2) + (n'^2 + n_f^2)/N \right], \tag{II.13}$$

respectively.

The final value of the magnetization in the equilibrium state can be calculated by maximizing the entropy subject to the condition that the energy must be conserved. This yields

$$M_f = \tfrac{1}{2} \gamma \hbar n_f = M_0 [1 + (B_{loc}/B_1)^2]^{-1}. \tag{II.14}$$

To summarize, for $B_1 \gg B_{loc}$ the spin system is unable to provide the entire energy for the decay of the transverse magnetization predicted by the Bloch equations. The state described by Eq. (II.14) is an equilibrium state and can be disturbed only by external interactions, for instance the spin lattice interaction.

Equation (II.14) has been verified experimentally by Slichter and Holton [36]. It is therefore reasonable to characterize the spin locked I spin system by a spin temperature in the rotating frame. Similarly, it can be shown that the S spin system can also be considered as a thermal reservoir of a certain temperature. Moreover, the rules of thermodynamics can be applied in the rotating frame to calculate the thermal properties of the systems, e.g. the heat capacities or the time constant for the approach to the equilibrium temperature, which corresponds to thermal conductivity of the rod in the model experiment. These quantities can be measured in a DNMR experiment. DNMR thus provides the possibility to check the concept of spin temperature in the rotating frame. This will be discussed in part V.

III. Experimental Procedure

The DNMR experiment combines two major steps: First, the I spin system is brought into a highly ordered state. Second, the S spin system must be transformed into a state of maximum disorder and coupled to the I spin system. The disorder of the S spin system must be maintained continuously. These two steps can be realized in various ways and are first discussed separately. Since the detection of the remaining order depends on the way in which the ordered state of the I spin system was prepared, it is discussed together with the preparation of the ordered state. At the end of this part a description of a typical experimental setup of a DNMR spectrometer is given.

III.1. Spin Locking and Detection

III.1.A *Spin Locking by a 90° Pulse and Subsequent Sudden 90° Phase Shift.*

Spin locking can be accomplished by use of the technique which has been discussed in section II.3. For the sake of completeness it shall be summarized briefly, referring to Fig. 4. A 90° pulse (B_{1I} antiparallel to y') flips the I spin magnetization from the z direction into the x' direction of the rotating frame. Subsequently the phase of B_{1I} is shifted by 90° rapidly as compared to T_2. Thus M_I and B_{1I} are parallel, which means the I spins are spin locked and M_I will decay slowly because of interaction with the lattice with a time constant $T_{1\varrho}$, comparable to T_1. At the end of the spin locking the remaining order of the I spin system can be monitored by measuring the free induction decay (FID) signal after switching off B_{1I} suddenly.

The order of the I spin system is conserved, if B_{1I} is turned off adiabatically, i.e. if the condition $\partial B_{1I}/\partial t \ll 1/\gamma T_2^2$ is satisfied [36]. It is possible to perform a DNMR experiment using an I spin system which in this way is adiabatically demagnetized in the rotating frame. However, the double resonance condition Eq. (II.8) has to be replaced by a modified condition

$$\gamma_S B_{1S} \approx \gamma_I B_{\text{loc}} . \tag{II.8a}$$

Using this procedure the requirement of two strong rf fields during the entire DNMR experiment can be avoided. The remaining I spin order can be measured, if B_{1I} is turned on again adiabatically. The transverse I spin magnetization is then restored and can be monitored by observing the FID signal.

III.1.B *Spin Locking by an Adiabatic Approach to Resonance*

Instead of the relatively complicated pulse sequence, described in the previous section, Lurie and Slichter [13] have used alternatively an adiabatic approach to the resonance in order to obtain spin locking of the I spins. Suppose, the external magnetic field B_0 is, at a time $t = t_0$, adjusted to a value which exceeds the resonance field strength, $B_{00} = \omega_I/\gamma_I$, by an amount of several times the width of the I spin resonance line. If one now turns on B_{1I}, the effective field is still essentially parallel to B_0. If now B_0 is slowly decreased to B_{00}, the magnetization remains parallel to the effective field and will finally point parallel to B_{1I} in the rotating frame. Thus spin locking is accomplished by a simple adiabatic approach to the resonance [32], provided the following conditions are satisfied:

$$dB_0/dt \ll \gamma_I B_{1I}^2 , \tag{III.1}$$

and

$$dB_0/dt \gg (\gamma_I T_1 T_2)^{-1} (1 + \gamma_I^2 B_{1I}^2 T_1 T_2)^{1/2} . \tag{III.2}$$

The remaining order of the I spin system at the end of the spin locking can be measured using the same technique as in section III.1.A.

III.1.C *Transformation of Zeeman Order in the Laboratory Frame into Dipolar Order in the Rotating Frame*

Spin locking transforms the nuclear polarization in the laboratory frame into a metastable magnetization transverse to the static external magnetic field B_0. As mentioned previously, the order of the spin system may be conserved even in the demagnetized state, for instance after an adiabatic demagnetization in the rotating frame [36]. Zeeman order, however, can also be transformed into dipolar order by applying a sequence of two rf pulses [37], which shall be discussed here in a simplified way, using Fig. 7.

At the time $t = 0$ a first 90° pulse (B_{1I} pointing along the $-y'$-direction) is applied to the spin system, which originally was polarized in B_0. In this way the magnetization vector is turned to the $+x'$-direction (Fig. 7A), yet because of fluctuations of B_{loc} the magnetization fans up in the $x'y'$ plane. After a time of the order of T_2, spins at places where B_{loc} is parallel to B_0 will preferentially point into the $-y'$-direction, those with B_{loc}

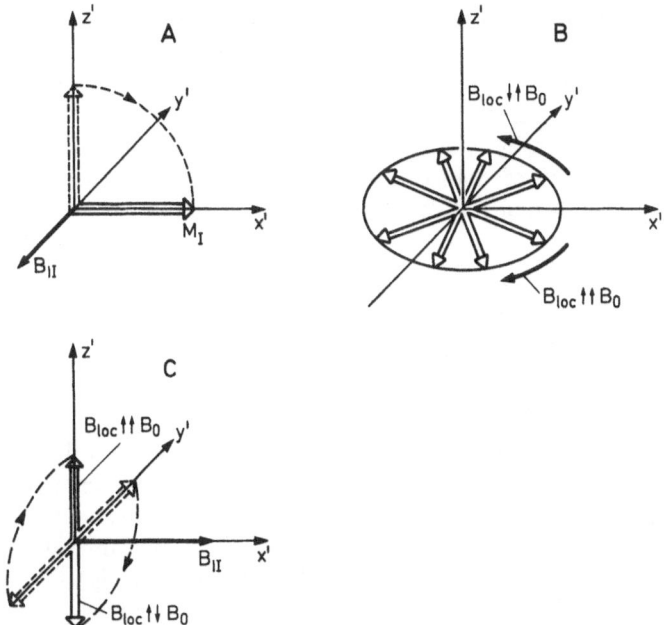

Fig. 7. Transformation of Zeeman order into dipolar order. A A 90° pulse turns the magnetization vector from the z-direction to the x'-direction of the rotating frame. B Fast and slow spins (B_{loc} parallel or antiparallel to B_0) precess in opposite directions in the rotating frame. After a time of the order of T_2, the fast spins point preferentially into the $- y'$-direction, the slow spins into the $+ y'$-direction. C A second 90° pulse brings the fast spins into the $+ z$-direction, the slow ones into the $- z$-direction. Both spin species are thus aligned in their respective local fields

antiparallel to B_0 into the $+ y'$-direction (Fig. 7 B). If now a second 90° pulse (B_{1I} parallel to x') is applied, spins parallel to the $+ y'$-axis are turned to the $- z$-direction, those parallel to $- y'$ to the $+ z$-direction (Fig. 7 C). Thus, by applying this pulse sequence, dipolar order has been created, since the spins now point preferentially along their respective local fields.

The dipolar order can be measured by applying an additional 90° pulse which brings the magnetization vectors along the $\pm z$ directions back into the $x'y'$ plane. The transverse magnetization will then be recovered similarly to a spin echo experiment. It is obvious that the description given in the previous paragraphs only makes this technique plausible. A detailed theory of it is given in Ref. [37].

If the pulse sequence discussed in the previous paragraphs is used, only a fraction of the original Zeeman order is transformed into dipolar order and can be recovered as a macroscopic magnetization. Jeener and

Broekaert [37] have pointed out that the transformations are most efficient, if a $90°_{y'} - 45°_{x'}$ pulse sequence is used, which for the detection is followed by a $45°_{x'}$ pulse. However, even under these optimum conditions, only $\frac{1}{3}$ of the original magnetization is regained. Yet the resulting loss in signal intensity is at least partially compensated for, because the transformation of Zeeman order into dipolar order can be performed relatively easily using standard pulse spectrometers.

III.1.D *Continuous Spin Locking*

The methods discussed in sections III.1.A through C yield a quasi equilibrium state, which is destroyed because of the interaction with the lattice in times of the order of T_1. It is. however, also possible to achieve continuous spin locking. Redfield [15] has pointed out that the magnetization of a system of tightly coupled spins, which are exposed to a strong rf field near resonance, precesses in phase with B_{1I}, provided the saturation condition $\gamma_I^2 B_{1I}^2 T_1 T_2 \gg 1$ is fulfilled. The equilibrium value of the x' component of the magnetization in the rotating frame is

$$M_{x'I} = \frac{M_{0I}\gamma_I B_{1I}(\omega_I - \omega_{0I})}{(\omega_I - \omega_{0I})^2 + \gamma_I^2(B_{1I}^2 + 2B_{loc}^2)}, \tag{III.3}$$

where M_{0I} is the equilibrium magnetization in the static field B_0,

$$M_{0I} = \chi_{0I} B_0 / \mu_0. \tag{III.4}$$

χ_{0I} is the static susceptibility, $\omega_{0I} = \gamma_I B_0$ the I spin Larmor frequency, and μ_0 the permeability of the vacuum. $M_{x'}$ has the extreme values

$$M_{x'I_{min}^{max}} = \pm \frac{1}{2} M_{0I} B_{1I}(B_{1I}^2 + 2B_{loc}^2)^{-1/2}, \tag{III.5}$$

which for $B_{1I} \gg B_{loc}$ reaches the asymptotic value

$$M_{x'I\,max,\,max} = M_0/2. \tag{III.6}$$

This compares to the maximum obtainable transverse magnetization in liquids at stationary conditions [32]

$$M_{x',\,max,\,max}^{liqu.} = \frac{1}{2} M_0 \sqrt{T_2/T_1}. \tag{III.7}$$

Thus in solids $M_{x'}$ can be several orders of magnitude larger than one would expect according to Bloch's equations. After a sudden change $M_{x'I}$ reaches its steady state value in times of the order of T_1.

Once again this transverse magnetization can be destroyed by irradiation of an audio field perpendicular to B_{1I} and $M_{x'I}$, if the resonance condition $\omega_a = \gamma_I B_{\text{eff}I}$ is fulfilled. In fact, Redfield has performed his famous rotary saturation experiment using this spin locking technique. Similarly a stationary DNMR experiment can be carried out [38]. According to Eq. (III.5) an rf field at a level far above saturation, whose frequency coincides with the wing of the I spin resonance line, creates a large transverse I spin magnetization. A second rf field B_{1S} creates disorder in the S spin system and by matching the double resonance condition Eq. (II.8) both systems can be coupled. There is a competition between the flux of disorder from the S spin to the I spin system and the reestablishment of the originally ordered state of the I spin system. But since the latter process takes place in times of the order of T_1, the I spin system integrates over the total flux of disorder from the S spin system. Consequently a large amount of disorder can be transferred even by a small number of S spins, if they are kept in disorder continuously.

For the detection of the remaining I spin magnetization Jones and Hartmann used, instead of a continuous B_{1I} field, a coherent rf pulse train, whose repetition time was short compared to T_2. Thus, the I spins were exposed only to the central frequency of the rf field and not to the sidebands. The receiver monitored the I spin magnetization only during the period while B_{1I} was switched off. In this way rf leakage problems could be avoided. This method of continuous spin locking and detection was more convenient and reduced stability problems. However, the B_{1I} modulation caused the I spin magnetization to be modulated. Therefore, DNMR signals were observed whenever the frequency of B_{1S} coincided with one of the I magnetization sidebands. These interfering resonances could be identified since they shifted in frequency when the modulation frequency was shifted.

III.2. Destruction of Order in the S Spin System

III.2.A *Sudden Switching of a Resonating* rf *Field* B_{1S}

If an rf field B_{1S}, which exactly fulfills the resonance condition $\omega_S = \gamma_S B_0$, is turned on suddenly, order which is stored in the S spin system will be destroyed in times comparable to T_2 (see section II.2). If, however, the double resonance condition Eq. (II.8) is satisfied the I and the S spin systems are coupled together. The resulting exchange of order causes spin locking of the S spins. An additional exchange of order can take place only if the S spin order is destroyed once again. This can be done most easily by switching off B_{1S} suddenly. The transverse magnetization then decays to zero in a time comparable to T_2 and the order of the S spin system is destroyed irreversibly. Following this decay, B_{1S} can again be turned on suddenly and the whole cycle can be repeated [13].

Hartmann and Hahn [12] used a similar but somewhat more efficient technique: B_{1S} is not switched off after equilibrium between the I and the S spin system is established, but rather its phase is shifted suddenly by 180°. The S spin magnetization is then antiparallel to B_{1S}, which is an extreme nonequilibrium state. Mutual $I-S$ spin flipping processes reestablish thermal equilibrium between both spin systems. This technique offers two advantages: First, the loss of I spin order is twice as large compared to the sudden switching technique, since M_S reverses its sign and does not only increase from zero to its equilibrium value. Second, there is a continuous exchange of order between the I and S spin system, whereas in the sudden switching technique the two systems are decoupled during the off periods of B_{1S}. Thus the method of shifting the phase of B_{1S} suddenly should be about four times as efficient as the sudden switching.

III.2.B *Continuous Saturation of the S Spin System*

The sudden switching of the rf field as discussed in the previous section corresponds in the model experiment (section II.1) to the repeated heating of the S system followed by the heat exchange. In a DNMR experiment the other alternative can be realized, too: The S spins can be kept continuously in a highly disordered state using rotary saturation and, at the same time, the exchange of order between both spin systems can be maintained. Suppose, one would modulate B_0 with an audio field $B_a = B_{0a} \cos \omega_a t$, whose frequency satisfies the resonance condition $\omega_a = \gamma_S B_{1S}$. This would cause a destruction of the S spin magnetization (see section II.4). However, since the double resonance condition $\gamma_S B_{1S} = \gamma_I B_{1I}$ has to be observed, the I spin magnetization would be destroyed at the same time. But, one can modulate the effective field for the S spins without disturbing the I spins simply by frequency modulating B_{1S} in the following way: $B_{1S}(t) = 2B_{1S} \cos(\omega_S - \Delta\omega_S \cdot \cos\omega_a t) \cdot t$. This procedure has been particularly helpful in zero field quadrupole double resonance experiments (see section IV.3). Rotary saturation in these experiments can help to identify the S spin resonance signals.

Rotary saturation can also be accomplished by modulating the external field without disturbing the I spin system, if it is totally demagnetized in the rotating frame (see section III.1.A and C). Since the I spins are not exposed to an rf field, they do not respond to an audio modulation of B_0 and their internal degree of order is not influenced by it.

III.3. A Typical DNMR Spectrometer

Usually a conventional NMR pulse spectrometer is the basic unit of a DNMR spectrometer, except that two different rf fields must be applied to the sample. The spectrometers which have been described in the literature [12, 13, 17, 22] are different only in minor details, even though different methods of spin locking of the I system and heating of the S

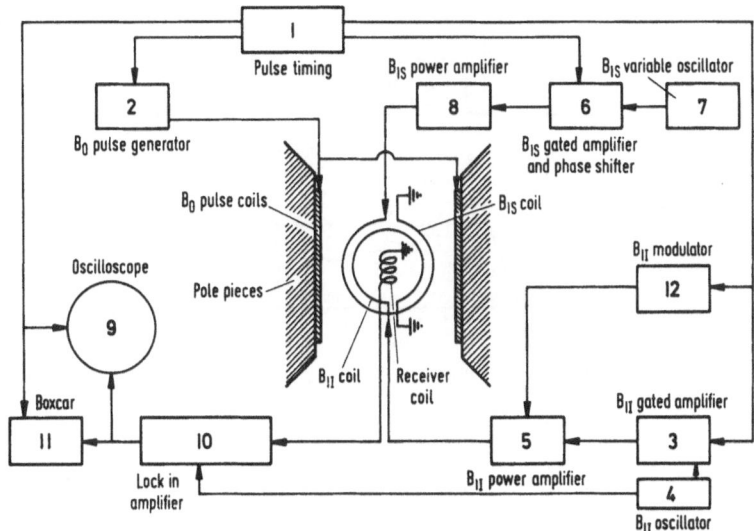

Fig. 8. Block diagram of a typical DNMR spectrometer

system have been used. In this section a typical spectrometer shall be discussed using the block diagram of Fig. 8. The pulse sequence, which is mainly determined by T_1, and the duration of B_{1S}, including the exact synchronization of the individual pulses, are controlled by a central clock. Before starting the experiment, B_0 is adjusted accurately to its resonant value, $B_{00} = \omega_I / \gamma_I$. When the I spin system has reached its equilibrium magnetization the central clock triggers an adiabatic increase of B_0 of approximately 20 Gauss (pulse generator 2). At the maximum deflection of B_0 the B_{1I} pulse amplifier (3) and the B_{1I} power amplifier (5) are activated. The frequency of B_{1I} is controlled by a quartz oscillator (4). Now B_0 returns to the resonant value B_{00}, whereby the I spin system is spin locked (see section III.1.B). Subsequently the B_{1S} pulse amplifier (6) is activated. It is fed by an oscillator of variable frequency (7). The B_{1S} power amplifier (8) exposes the sample suddenly to the second rf

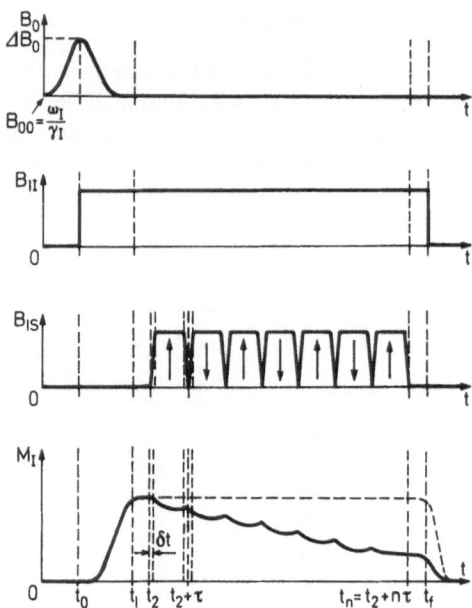

Fig. 9a. Timing of a DNMR experiment

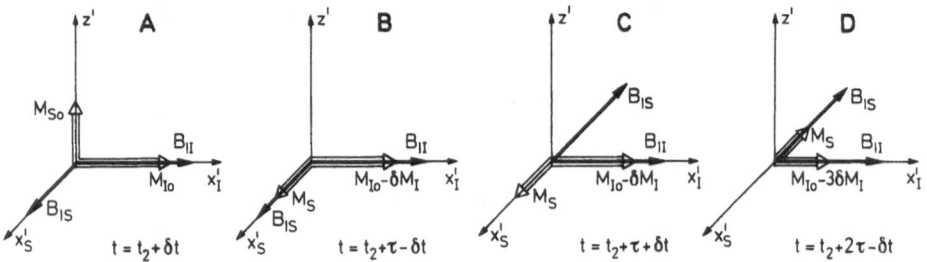

Fig. 9b. Magnetization vectors and effective fields in the rotating frame

The spin system is polarized in a magnetic field, $B_{00} = \omega_I/\gamma_I$. At the time $t = 0$, B_0 is increased adiabatically by an amount of $\Delta B_0 \gg B_{loc}$. B_{1I} is switched on suddenly at the maximum deflection of B_0. Now B_0 returns adiabatically to its resonant value, thus causing spin locking of the I spin system. At the time, $t = t_2$, B_{1S} is switched on within a period, $\delta t \ll T_2$, leading to the situation depicted in part A of Fig. 9b. Cross relaxation between the I and S spin systems decreases the I magnetization by δM_I and causes spin locking of the S system (part B of Fig. 9b). At the end of the cross relaxation ($t = t_2 + \tau$) the phase of B_{1S} is suddenly shifted by 180°. Now M_S and B_{1S} are antiparallel (part C of Fig. 9b). Cross relaxation now decreases M_I again by an amount of $2\delta M_I$ and gives rise to a spin locking of the S system (part D of Fig. 9b). This process can be repeated n times, which results in a cummulative destruction of the transverse I magnetization, M_I

field and creates disorder in the S spin system, if $\omega_S = \gamma_S B_{00}$. This disorder is maintained continuously using periodic 180° phase shifts of B_{1S} (modulator 6). At the end of the DNMR experiment the central clock switches off B_{1S} first, and a few msec later B_{1I}. At the same time it triggers the x deflection of an oscilloscope (9), which records the I spin FID signal, amplified by the lock in amplifier (10). A digital boxcar integrator (11) simplifies the evaluation of the data and simultaneously improves the signal to noise ratio, since it can average over several FID signals.

By turning off B_{1I} adiabatically the I spin system can easily be demagnetized in the rotating frame. The modulator (12) provides the possibility for such an adiabatic demagnetization and remagnetization. Fig. 9a illustrates the pulse sequence of a DNMR experiment. In Fig. 9b the magnetization vectors and effective fields in the doubly rotating frame are shown schematically.

IV. Application of DNMR in Solid-State Physics

IV.1. Introduction

In the previous chapters exclusively spin systems have been discussed whose properties are determined by magnetic interactions (Zeeman or dipolar, respectively). DNMR, however, is not only restricted to these systems, but can be extended to systems which are exposed to additional interactions, for instance quadrupole interaction. Quadrupole interactions have been studied in detail by use of conventional NMR technique [39–45]. These measurements yield information about the electronic structure of molecules and solids. But, if the concentration of the nuclei under consideration is low, the sensitivity has to be improved. This can be accomplished using DNMR techniques. For instance point defects in solids at concentrations as low as 1% to 10^{-4}% have been investigated. The results of these studies shall be presented in this section, and a slight modification of DNMR, the so called zero field quadrupole double resonance (NQDR), shall be discussed. These results have been mainly obtained in metals and ionic crystals, doped with a small amount of impurities. They yield valuable information about the structure and electronic properties of these materials.

IV.2. Preliminary Remarks Concerning Quadrupole Interaction

Quadrupole interaction and quadrupole spectroscopy have been discussed in great detail in excellent monographies [39, 40]. In this section

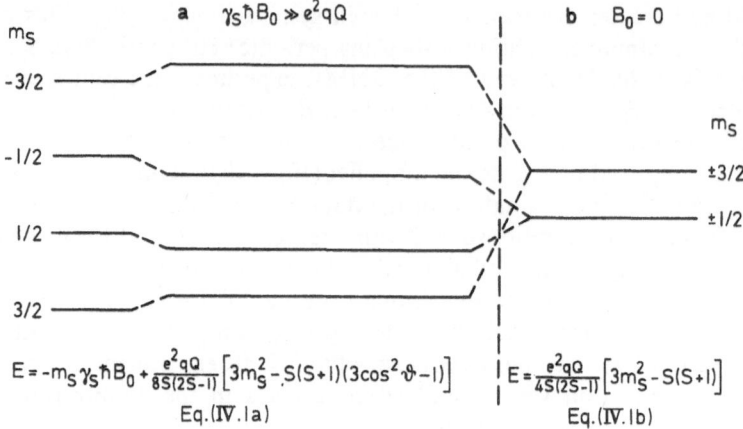

$$E = -m_S \gamma_S \hbar B_0 + \frac{e^2 qQ}{8S(2S-1)}\left[3m_S^2 - S(S+1)(3\cos^2\vartheta - 1)\right]$$

Eq. (IV. 1a)

$$E = \frac{e^2 qQ}{4S(2S-1)}\left[3m_S^2 - S(S+1)\right]$$

Eq. (IV. 1b)

Fig. 10. Energy levels of a spin $S = \frac{3}{2}$ without quadrupole effects (left), in a strong magnetic field and a weak, axially symmetric field gradient (center), and in an axially symmetric electric field gradient at zero magnetic field (right)

it is necessary to recall only a few of the most important properties. We will restrict ourselves to the simplest case of an axial symmetric electric field gradient (EFG), primarily.

Suppose, the quadrupole interaction is weak compared to the Zeeman interaction. The originally equidistant Zeeman levels of a spin $S = \frac{3}{2}$ are shifted as indicated in the center of Fig. 10, where we have used the abbreviation

$$eq = \sum_j (e_j/4\pi\varepsilon_0 r_{ij}^3)\,(3\cos^2\varphi_{ij} - 1), \tag{IV.2}$$

φ_{ij} is the angle between the axis of symmetry and the vector r_{ij} from the nucleus i to be studied to the charge e_j in the lattice. eQ is the nuclear quadrupole moment, and ϑ the angle between the symmetry axis and the external magnetic field. The energy difference between the levels $|m_S\rangle$ and $|m_S - 1\rangle$ of the nuclear spin is given by first order perturbation theory

$$\Delta E_{m_S \to m_S - 1} = \gamma_S \hbar B_0 + (2m_S - 1)\frac{3e^2 qQ}{8S(2S-1)}(3\cos^2\vartheta - 1). \tag{IV.3}$$

Denote the average value of the quantum numbers involved

$$m_q = \tfrac{1}{2}(m_S' + m_S''). \tag{IV.4}$$

Then Eq. (IV.3) reads

$$\Delta E_{mq} = \gamma_S \hbar B_0 + m_q \frac{3e^2 qQ}{4S(2S-1)}(3\cos^2\vartheta - 1). \tag{IV.5}$$

Thus the NMR line splits into an equidistant triplet, whose central component ($m_S = \frac{1}{2} \leftrightarrow m_S = -\frac{1}{2}$) is not shifted in first order.

At points around which the symmetry is cubic, $q = 0$, and there is no quadrupole interaction. But if the cubic symmetry is destroyed by an imperfection, e.g. an impurity atom, this causes a quadrupole splitting of the NMR line. The measurement of this splitting yields information about lattice distortions and charge distributions around the imperfection.

Consider, for example, aluminum ($I = \frac{5}{2}$) which is doped with impurity atoms like Zn or Mg at very low concentrations. For nuclei close to the impurity the transitions $m_q = 1$ and $m_q = 2$ (satellites) are shifted so strongly that they don't contribute to the resonance signal of the unperturbed nuclei any more. By measuring the number of nuclei, which are "wiped out" by an impurity atom, Rowland [41] has shown that quadrupole interaction in aluminum is a long range interaction. However, a direct measurement of the field gradients in the vicinity of the impurity turned out to be very difficult. Because of the small skin depth it is difficult to study metals as single crystals. Therefore a so called powder pattern is observed, whose interpretation is more difficult than that of the simple multiplet according to Eq. (IV.5). Drain [46] has successfully studied such imperfections in metals using conventional NMR. But the zero field quadrupole resonance (NQDR), invented by Redfield [14], has proven to be much more sensitive and versatile. It shall be discussed in the next section.

IV.3. Zero Field Quadrupole Double Resonance (NQDR)

IV.3.A *Principle*

The application of an external magnetic field to a powder sample results in an undesirable broadening of the NMR line, as discussed in the previous section. It can be avoided by applying zero field quadrupole spectroscopy. One then observes transitions between the levels, shown in the right hand part of Fig. 10. The transition frequency is the same for all nuclei with the same quadrupole interaction, independent of their orientation. However, if the concentration of the nuclei under study is small, the sensitivity of a straight forward quadrupole resonance experiment may be too low. In NQDR experiments one uses a magnetic field sequence which combines the advantages of pure quadrupole spectroscopy with the sensitivity of DNMR. Fig. 11 illustrates the typical pulse sequence of such an experiment: First, the sample is polarized in a high magnetic field B_{00}. Second, it is demagnetized adiabatically by turning off B_0

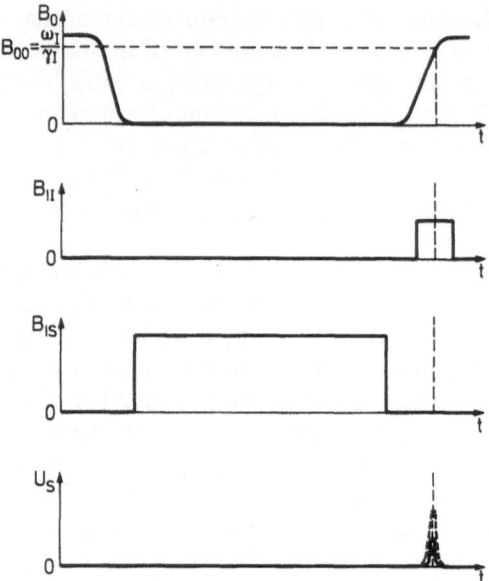

Fig. 11. Zero field quadrupole double resonance (NQDR). The sample is demagnetized adiabatically from a high magnetic field B_0. In the demagnetized state it is exposed to the search field B_{1S}, which induces nuclear transitions, if its frequency ω_S matches the quadrupole frequencies of the S spins. Cross relaxation between the I and S systems causes a loss of order within the I system. It can be detected by monitoring the I spin signal U_S in an adiabatic rapid passage experiment

rapidly compared to T_1, but slowly compared to T_2. The I spin temperature is then

$$\theta_{I0} = \theta_l B_{\text{loc}} / B_{00} \tag{IV.6}$$

θ_l is the lattice temperature, B_{loc} the local magnetic field (see Section II.3, and Eqs. (V.15) and (V.17), respectively.) If the sample is remagnetized in a time short compared to T_1 nearly the entire initial I spin magnetization will be recovered. But the S spin system will be heated, if an rf field is irradiated, which satisfies the condition

$$\nu_S = \frac{3e^2 qQ}{2S(2S-1)h} m_q, \tag{IV.7}$$

during the period of demagnetization. The S spin system can pass on the absorbed energy to the I spin system, if a condition similar to Eq. (II.8) is met:

$$\sqrt{S(S+1) - \tfrac{1}{4}(4m_q^2 - 1)}\, B_{1S} \approx \gamma_I B_{\text{loc}}. \tag{IV.8}$$

This causes a rise in the I spin temperature, which in turn reflects itself in a decrease of the I spin NMR signal after the remagnetization. The I spin NMR signal usually is observed by means of the technique of adiabatic rapid passage.

For frequencies less than about $v_S < 150$ kHz an additional modulation of B_{1S} is not indispensible [20, 26, 27, 34], but the sensitivity of the method increases, if B_{1S} is modulated properly. At frequencies higher than $v_S \approx 150$ kHz it is necessary to modulate B_{1S}, e.g. by 180° phase shifts or by frequency modulation similar to the rotary saturation. In addition this might help to identify the S spins to be studied. This method will be discussed in more detail in Section IV.3.C.

IV.3.B *Results Obtained on Atomic Defects in Metals*

In Fig. 12 two typical NQDR spectra are shown, which were obtained with powdered samples of 99.9999 % aluminum (curve a) and aluminum doped with 0.4 % Mg (curve b), respectively [27]. The data were taken at 1.4 °K, at which temperature the zero field spin lattice relaxation time of pure aluminium is about 0.5 sec. During the zero field period an unmodulated B_{1S} field of about 70 mG was irradiated for about 200 msec. The I spin signal remaining after the NQDR cycle, S_{DR}, was normalized to the signal S_0, which was obtained after the same pulse sequence but without B_{1S} irradiation. Several lines were observed: Line A results from the so called nonresonant spin absorption [44] ($\omega_S \approx \gamma_I B_{loc}$). The

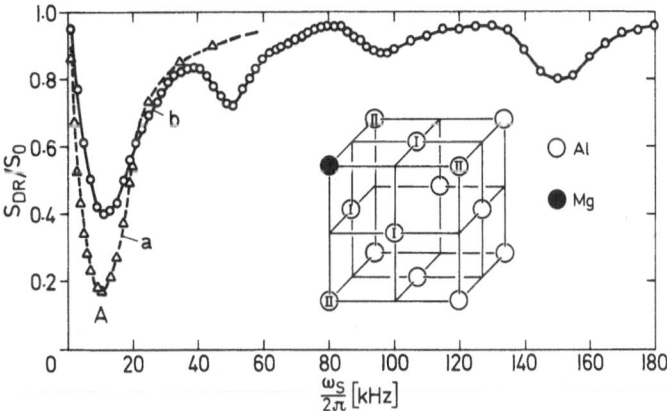

Fig. 12. NQDR spectra obtained in powdered samples of 99.9999 % aluminium (curve a), and aluminium doped with 0.4 % Mg (curve b) [27]. The unmodulated search field B_{1S} was irradiated for 0.5 sec at 1.4 °K. The I spin resonance signal remaining after the NQDR cycle, S_{DR}, was normalized to the signal S_0, which was obtained after the same pulse sequence but without B_{1S}

other three lines, and a fourth one which was observed using a modulated B_{1S} field, can be interpreted as quadrupole transitions of aluminum nuclei in the vicinity of the impurities.

The following section gives a simplified interpretation of these lines. An S spin of quadrupole moment Q in an inhomogeneous electric field is described by the following Hamiltonian [40]

$$\hat{\mathcal{H}}_Q = A[(3\hat{S}_Z^2 - \hat{S}^2) + \eta(\hat{S}_X^2 - \hat{S}_Y^2)] , \tag{IV.9}$$

where

$$A = \frac{e^2 q Q}{4S(2S-1)} = \frac{eQ \, \partial^2 V/\partial Z^2}{4S(2S-1)} , \tag{IV.9a}$$

and

$$\eta = \frac{\partial^2 V/\partial X^2 - \partial^2 V/\partial Y^2}{\partial^2 V/\partial Z^2} . \tag{IV.9b}$$

$\partial^2 V/\partial X^2$, $\partial^2 V/\partial Y^2$ and $\partial^2 V/\partial Z^2$ are the second derivatives of the electrostatic potential at the S spin site in the principal axes directions of the field gradient tensor, (X, Y and Z), respectively. For a spin $S = \frac{3}{2}$ this Hamiltonian can be diagonalized exactly which yields a pair of doubly degenerate energy levels

$$E_{\pm\frac{1}{2}} = -3A(1 + \tfrac{1}{3}\eta^2)^{1/2} ,$$
$$E_{\pm\frac{3}{2}} = 3A(1 + \tfrac{1}{3}\eta^2)^{1/2} , \tag{IV.10}$$

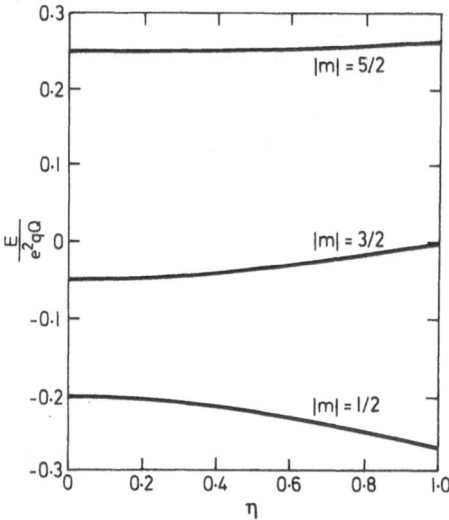

Fig. 13. Energy levels of a spin $S = \frac{5}{2}$ in zero magnetic field in an electric field gradient eq of asymmetry parameter η

Thus a single quadrupole resonance transition can be observed at a frequency

$$\omega_S = 6A\hbar^{-1}(1 + \tfrac{1}{3}\eta^2)^{1/2}, \tag{IV.11}$$

which does not allow to determine A and η separately. For $S > \tfrac{3}{2}$ the secular equation can be solved only numerically [45]. The result of such a solution is plotted in Fig. 13 for $0 \leq \eta \leq 1$ and for the case of $S = 5/2$ (which for instance applies to aluminum). In this case two quadrupole transitions can be observed, the frequency ratio of which may allow to determine A and η separately. In the example of Fig. 12 the lines at 50 kHz and 100 kHz belong to Al nuclei at sites with axial symmetry, those at 155 kHz and 295 kHz to sites with a nonaxial symmetric field gradient ($\eta = 0.20 \pm 0.05$).

The crystal structure of aluminum is fcc. Thus the nearest neighbors to a substitutional impurity (type I in Fig. 12) are at sites with nonaxial symmetry, whereas the $\langle 200 \rangle$ neighbors (type II in Fig. 12) are at sites with axial symmetry. Thus the pairs of lines can be associated with them, yielding

$$eq_{\mathrm{I}} = 2.66 \cdot 10^{16} \ \text{V/cm}^2 \text{ for nuclei in } \langle 110 \rangle \text{ positions},$$

and

$$eq_{\mathrm{II}} = 9.25 \cdot 10^{15} \ \text{V/cm}^2 \text{ for nuclei in } \langle 200 \rangle \text{ positions}.$$

Table IV.1. Electric field gradients at the first two shells around an impurity atom in aluminum. The values are given in units of 10^{16} V/cm^2[a]

Charge difference	Host	Impurity	Shell I $eq\sqrt{1 + \tfrac{1}{3}\eta^2}$		Shell II $eq\sqrt{1 + \tfrac{1}{3}\eta^2}$		Ref.
0	Cu	Ag	3.26		2.50		[14]
+1		Zn	> 26		10.4		[14]
			eq	η	eq	η	
−2	Al	Ag	3.1	0.36 ± 0.05	0.75	0	[27]
−1		Mg	2.66	0.20 ± 0.05	0.92	0	[26, 46]
−1		Zn	2.59	0.37 ± 0.05	0.50	0	[20]
0		Ga[b]	3.23	0	0.41		[27]
0		In[b]	0.9		4.9	0	[27]
+1		Ge[b]	0.8		4.6	0	[27]
+1		Si[b]	1.10		4.0	0	[27]

[a] To obtain values in e.s.u. divide by 300, to obtain atomic units multiply with $0.514 \cdot 10^{-18}$.

[b] Originally [27] the field gradients were assigned to shell I and II in the opposite way. Recent results on Al:Ga support the assignment given here [47].

Table IV.1 provides a collection of field gradients around substitutional atoms in metals as available to date.

The first measurements of the wipe out numbers [41] already proved that the field gradients around impurity atoms in metals cannot be explained simply because of lattice distortions. On the contrary, the change in the charge distribution of the conduction electrons plays the most important role. A calculation of this charge distribution, based on the scattering of electrons near the Fermi surface [48, 49], yielded the result that there should exist a long range oscillation in the field gradients of the type

$$q(r) = K \, r^{-3} \cos(k_F r + \phi) \, . \tag{IV.12}$$

r is the distance from the impurity, and k_F the wave vector of the electrons at the Fermi surface. $q(r)$ is entirely determined by the knowledge of the Bloch functions and the phase shift ϕ, and could be calculated in principle. However, Blandin and Friedel have determined the amplitude K and the phase shift ϕ semiempirically by adjusting the calculated values of wipe out numbers and changes in resistivity to the experimental data. In this way they avoided uncertainties in the ionic polarizabilities and the Sternheimer antishielding factors in obtaining the field gradients in Al:Mg and Al:Zn.

Table IV.2. Comparison of experimental and theoretical values of the electric field gradients in the vicinity of atoms in aluminum

Sample	$eq(10^{16} \text{ V/cm}^2)$			
	Shell I		Shell II	
	Theoretical	Experimental	Theoretical	Experimental
Al:Mg	1.81	2.66	0.60	0.92
Al:Zn	1.12	2.59	0.37	0.50

In Table IV.2 the theoretical values of the field gradients are compared with the experimental data. The agreement is reasonable, but this fact should be virtually considered to be fortuitous, since the scattering concept is valid only for distances large compared to $1/k_F$. Even small adjustments of K and ϕ can change the agreement by a large amount.

Stimulated by the experimental determination of the field gradients around impurity atoms Fukai and Watanabe [51] have tried to take into account the overlap between conduction electrons and atomic core functions. Using a pseudopotential method they calculated the field gradients around all the impurities studied by Minier [27]. But the agreement is rather unsatisfactory, particularly the calculations yield axial symmetric field gradients only. Based on previous calculations of resistivities [52] they suggest that the Bloch wave character of the conduction electron wave functions must be accounted for in order to improve the theoretical results.

Drain [46] has pointed out that the Fermi level near the impurity is shifted, which causes a redistribution of the occupied states. Again, this can result in a considerable change of the field gradients. Based on the work of Watson and coworkers [53] he estimates the size of this effect. He concludes that it could be large enough to explain the discrepancies between theoretical and experimental results. However, until now it has not been possible to explain the experimentally determined electrical field gradients and their asymmetry in a quantitative way.

IV.3.C NQDR *in Ionic Crystals*

NQDR allows the investigation of powder samples, which makes it particularly useful for the study of impurities in metals. Impurities in ionic crystals can be studied in principal in single crystals. In spite of this, NQDR can be superior to high field DNMR, for instance if the impurity does not perturb the I spins only slightly. This will be discussed in Section IV.4.C. Substitutional ions like K^+ and Br^- in NaCl give rise only to small lattice distortions which should be relatively easy to study theoretically. This fact has stimulated strong efforts to analyze these perturbations experimentally.

Until now only field gradients next to substitutional ions in NaCl crystals have been studied. For this purpose the unperturbed ^{23}Na nuclei with an isotopic abundance of 100% and a relatively large γ of 7.076 $\cdot 10^3$ sec^{-1} G^{-1} have been used as I spins. Since the spin lattice relaxation time of the ^{23}Na nuclei amounts to approximately 75 sec at liquid nitrogene temperature, the quadrupole resonance frequency can be irradiated for a reasonable time. Quadrupole resonance lines of nuclei next to substitutional ions of concentrations as low as 10^{16} cm^{-3} can be observed. Fig. 14 shows as an example three spectra which where observed in NaCl crystals doped with different impurities. The top spectrum has been obtained with a pure NaCl crystal (Harshaw), the middle and bottom one with samples doped with 0.1% KCl and 0.2% NaBr, respectively. The amplitude of B_{1s} was about 2 G with periodical phase shifts at a rate of 20 kHz. There are several lines caused by perturbed ^{23}Na nuclei which shall be discussed in the next section.

Three questions have to be answered to identify the lines:

1. Which defect causes the resonance line?

2. Which nuclear species does the line belong to? (^{23}Na, as well as ^{35}Cl and ^{37}Cl, have a spin of $\frac{3}{2}$ and non-vanishing quadrupole moments.)

3. Which lattice positions are occupied by the respective nuclei?

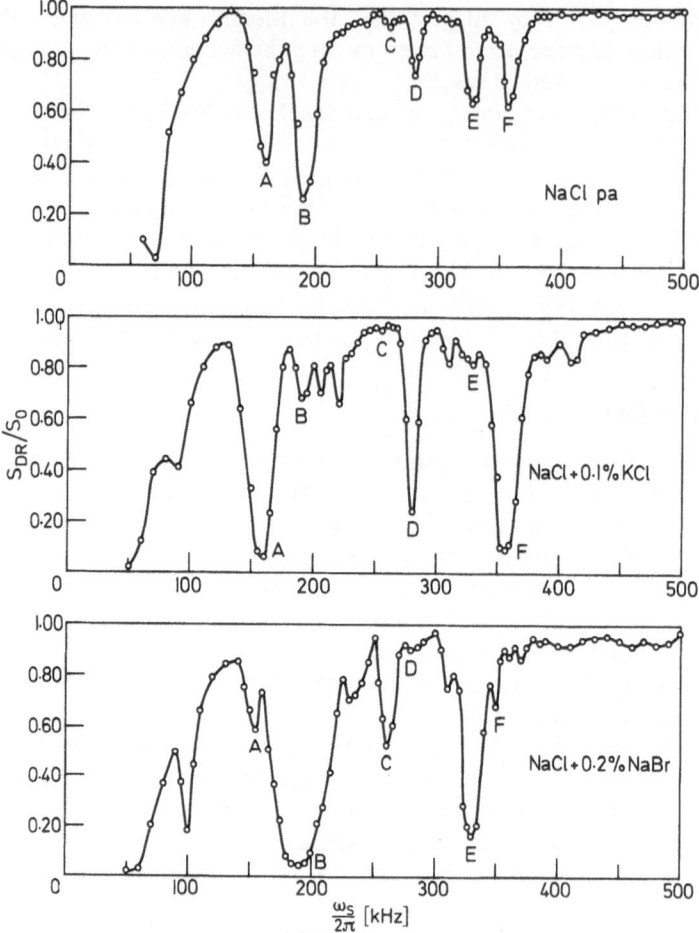

Fig. 14. NQDR in NaCl compiled from Ref. [21]. Top: Undoped crystal (Harshaw). Center: NaCl doped with 0.1 % KCl. Bottom: NaCl doped with 0.2 % NaBr

The first question can be answered by doping the crystals. Fig. 14, for instance, shows that the lines A, D and F are caused by nuclei next to a K^+ impurity, whereas the lines B, C and E result from Br^- impurities. Two additional lines, at 945 kHz and 1197 kHz, are not shown in Fig. 14. They were observed in samples doped with K^+.

It is more difficult to identify the perturbed nuclei. If there exist different isotopes of the same element with nonvanishing quadrupole moments, for instance ^{35}Cl and ^{37}Cl, the frequency of the corresponding lines have the ratio of the quadrupole moments of these nuclei. However, only in special cases it is possible to identify the defects definitely because of this frequency ratio.

The S spins can also be identified by use of rotary saturation, as mentioned in Section IV.3.A [21]. This method shall be indicated briefly for the case of an axial symmetric field gradient. Suppose, the z-axis coincides with the direction of the field gradient and B_{1S} is polarized parallel to the x-axis,

$$B_{1S}(t) = 2B_{1S} \cos(\omega_S t) \cdot x. \tag{IV.13}$$

Then the time dependent part of the Hamiltonian is

$$\hat{\mathscr{H}}_{1S}(t) = -\gamma_S \hbar B_{1S}(\hat{S}_+ + \hat{S}_-) \cos(\omega_S t). \tag{IV.14}$$

It causes transitions between the quadrupole energy levels $|m_S\rangle = |\pm\frac{1}{2}\rangle$ and $|m_S\rangle = |\pm\frac{3}{2}\rangle$ similar to the system of spins $S = \frac{1}{2}$ in a high magnetic field. However, both rotating components of B_{1S} are important in this case. The proper transformation of this Hamiltonian into a rotating frame has been given by Leppelmeier and Hahn [54]. It involves the concept of two rotating frames precessing in opposite directions, one for the transitions $m_S = \frac{1}{2} \leftrightarrow \frac{3}{2}$ and one for the transitions $m_S = -\frac{1}{2} \leftrightarrow -\frac{3}{2}$. In this system the eigenvalues are

$$E_{1S} = \pm\frac{1}{2}\sqrt{3}\,\gamma_S \hbar B_{1S}, \tag{IV.15}$$

if Eq. (IV.11) is fulfilled exactly and if B_{1S} is perpendicular to the field gradient. Both eigenvalues are doubly degenerate. If B_{1S} is not perpendicular to the field gradient, only the component perpendicular to it is effective. Analogous to rotary saturation (Section II.4) transitions between these levels can be induced by frequency modulation of B_{1S},

$$\omega_S(t) = \omega_{S0} + \Delta\omega_S \cdot \cos(\omega_a t). \tag{IV.16}$$

The resulting disorder in the S spin system can be transferred to the I spin system. This process will be most effective, if the resonance condition

$$\omega_a = \sqrt{3}\,\gamma_S B_{1S\perp}, \tag{IV.17}$$

is satisfied. Since B_{1S} and ω_a are measurable quantities γ_S can be determined, and because only $B_{1S\perp}$ is effective, the orientation of the field gradient relative to the crystal axis can be defined. In this way the NQDR lines have been associated with the defects compiled in Table IV.3. Until now, this method has been applied only for nuclei with a spin $S = \frac{3}{2}$ and for axial symmetric field gradients.

Table IV.3. NQDR lines observed in NaCl with K$^+$ and Br$^-$ impurities

$\omega_S/2\pi$(kHz)	Impurity	Resonant nucleus	Position relative to impurity
158	K$^+$	^{23}Na	$\langle 112 \rangle$
191.5	Br$^-$	^{23}Na	?
205	?	^{23}Na	?
210	?	^{23}Na	?
220	K$^+$	^{37}Cl	$\langle 111 \rangle$
259	Br$^-$	^{37}Cl	?
280	K$^+$	^{35}Cl	$\langle 111 \rangle$
328.5	Br$^-$	^{35}Cl	?
358	K$^+$	^{23}Na	$\langle 110 \rangle$
580	?	^{23}Na	?
944.5	K$^+$	^{37}Cl	$\langle 100 \rangle$
1197	K$^+$	^{35}Cl	$\langle 100 \rangle$

Summarizing, NQDR makes it possible to measure quadrupole splittings of nuclei at very low concentrations. The principal advantage of this method is that powder samples can be studied. However, the identification of the observed nuclei is more difficult than with high field methods. Rotary saturation can help to avoid this difficulty. However, even with this method, only a few nuclei next to K$^+$ impurities in NaCl could be identified definitely. Defects next to Br$^-$ impurities could not be associated with the observed lines without doubt.

IV.4. High Field Double Resonance

IV.4.A Quadrupole Shift in a High Magnetic Field

The interpretation of NQDR spectra proved to be difficult in many cases. Samples which can be investigated as single crystals can be studied more easily by using the high field DNMR method. The frequency of the transition $m_S = -\frac{1}{2} \leftrightarrow m_S = +\frac{1}{2}$ yields γ_S. In this way the nucleus under consideration can be identified. The angular dependence of the spectra gives information about its position relative to the impurity, and about the symmetry of the field gradient. In many cases the quadrupole interaction, Eq. (IV.9) is only a small perturbation to the Zeeman interaction

$$\hat{\mathscr{H}}_{ZS} = -\gamma_S \hbar B_0 \hat{S}_z . \tag{IV.18}$$

The direction of B_0 defines the z-axis of the laboratory frame and the eigenfunctions in zeroth order. In order to study the influence of $\hat{\mathscr{H}}_Q$ it must be transformed into the laboratory frame. For this transformation the quadrupole Hamiltonian is first rotated around the Z-axis by an angle φ until the X-axis and the x-axis of the laboratory frame coincide. Subsequently, this coordinate system is rotated around the common axis by an angle ϑ until the Z-axis coincides with the z-axis. (ϑ therefore is the angle between the Z and the z-axes, φ the angle between the X and the x-axes.) First order perturbation theory with the transformed Hamiltonian $\mathscr{H}_Q^*$ yields the eigenvalues [55]

$$E_{mS}^{(1)} = -\gamma_S \hbar B_0 m_S + W_Q[3m_S^2 - S(S+1)],$$ (IV.19)

with

$$W_Q = \frac{e^2 qQ}{4S(2S-1)} \left[\tfrac{1}{2}(3\cos^2\vartheta - 1) - \tfrac{1}{2}\eta \sin^2\vartheta \cos^2\varphi\right].$$ (IV.20)

These energy shifts are indicated in the center of Fig. 10 for the case of an axial symmetric field gradient and for $S = \tfrac{3}{2}$. The originally single line splits into a triplet of frequencies

$$\nu_{mq} = (2\pi)^{-1}\gamma_S B_0 - 6W_Q m_q/h.$$ (IV.21)

The angular dependence of this triplet allows to determine the principal axes and eigenvalues of the quadrupole Hamiltonian. In addition the center line yields γ_S. Thus it is easy to identify the nucleus S under study.

Frequently first order perturbation theory is insufficient to fit the observed spectra. Second order perturbation theory is not difficult in principal, but somewhat tedious. For $S = \tfrac{3}{2}$ one obtains the following second order corrections:

$$\delta\nu_{+1}^{(2)} = \frac{3\pi}{4}\frac{e^4 q^2 Q^2}{h^2 \gamma_S B_0}\{\sin^2\vartheta \cos^2\vartheta(1 + \tfrac{2}{3}\eta \cos^2\varphi)$$
$$+ \tfrac{1}{9}\eta^2 \sin^2\vartheta(1 - \sin^2 2\varphi \sin^2\vartheta)\},$$ (IV.22)

$$\delta\nu_0^{(2)} = \frac{3\pi}{32}\frac{e^4 q^2 Q^2}{h^2 \gamma_S B_0}\{\sin^2\vartheta(1 - 9\cos^2\vartheta) - \tfrac{2}{3}\eta \cos^2\varphi \sin^2\vartheta(1 + \cos^2\vartheta)$$
$$+ \tfrac{1}{9}\eta^2(4 + 12\sin^2\vartheta - 9\cos^2\varphi \sin^4\vartheta)\},$$ (IV.23)

$$\delta\nu_{-1}^{(2)} = \delta\nu_{+1}^{(2)}.$$ (IV.24)

The subscripts 1, 0 and -1 are the respective average values of the eigenvalues of the two states involved in the transition as defined by Eq. (IV.4).

IV.4.B Results on Point Defects in Alkali Halides

High field DNMR has been particularly successful in the investigation of point defects in alkali fluorides. The ^{19}F nuclei offer themselves as

I spins because of their large gyromagnetic ratio. Furthermore, they have a spin $I = \frac{1}{2}$ and therefore no quadrupole moment. Thus their resonance frequency, $\omega_{I0} = \gamma_I B_0$, is not perturbed even in the neighborhood of the impurity. On the contrary the alkali ions next to an impurity are perturbed and can be considered as S spins. In the most common case of $S = \frac{3}{2}$ the resoance lines split into triplets, with the frequencies given by Eqs. (IV.21) or (IV.22–24).

Figure 15 shows a DNMR spectrum as measured by Hartland [23]. It was obtained using a KF sample doped with 0.02 % NaF. The external magnetic field was $B_0 = 7.490$ kG corresponding to a resonance frequency of the ^{19}F nuclei, $\omega_{I0} = 30.00$ MHz. Unperturbed ^{39}K nuclei at this field have a resonance frequency of 1.488 MHz. Irradiation of an rf field of the frequency ω_S close to this frequency destroyed all the order within the I spin system. In addition, several lines which could be associated with perturbed ^{39}K nuclei were observed. They were grouped nearly symmetrically around ω_S. In order to analyze the spectrum the crystal was rotated around the [001] axis, whereas B_0 remained in the (001) plane. In Fig. 16 the angular dependence of the observed lines is shown. α is the angle between B_0 and the [100] axis. The lines can be divided into two groups which can be fitted by Eqs. (IV.21–24) quite satisfactorily. Group A can be associated with ^{39}K nuclei in $\langle 110 \rangle$ positions relative to the impurity. The Z-axis is pointing towards the impurity, the Y-axis is a $\langle 011 \rangle$ axis perpendicular to the Z-axis and the X-axis is perpendicular to both of them. Evaluation of these spectra yields

$$e^2qQ/h = \pm 0.916 \text{ MHz}; \quad \eta = 0.607. \tag{IV.25}$$

Group B results from nuclei in $\langle 200 \rangle$ positions relative to the impurity with

$$e^2qQ/h = \pm 1.138 \text{ MHz}; \quad \eta = 0. \tag{IV.26}$$

The Z-axes of these nuclei are the [100], [010], and the [001] directions, respectively. Since the quadrupole splitting is comparable with the Larmor frequency the experimental points have been fitted by the exact numerical solution of the total Hamiltonian [23]. The line at 1.850 MHz in Fig. 15 results from unperturbed ^{40}K nuclei with an isotopic abundance of only 0.012 %. It illustrates the sensitivity of DNMR.

Similar results were obtained by Nelson [24, 25] in LiF crystals doped with NaF and KF, respectively. Recently, first order quadrupole splittings in the neighborhood of impurities in alkali halides have been studied successfully using conventional NMR techniques [56–58]. However, until now only data for axial symmetric field gradients have been published. In previous investigations [59–61] it was attempted

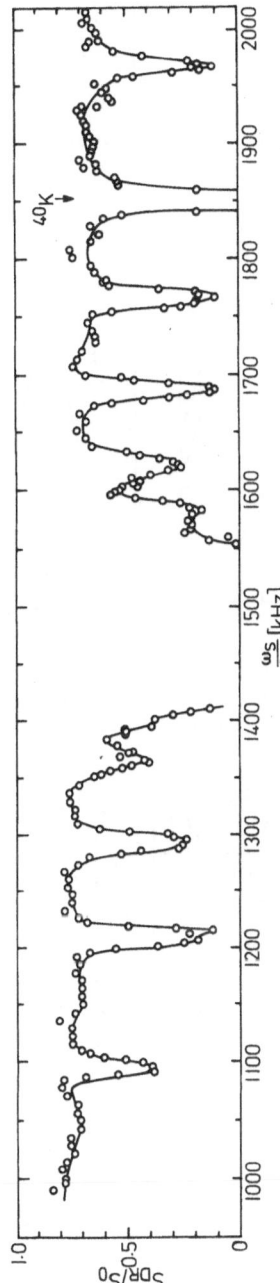

Fig. 15. DNMR spectrum obtained in KF doped with 0.02 % NaF [23]. B_0(7.490 kG) is in the (001)-plane and forms an angle of 22° with the [100]-axis

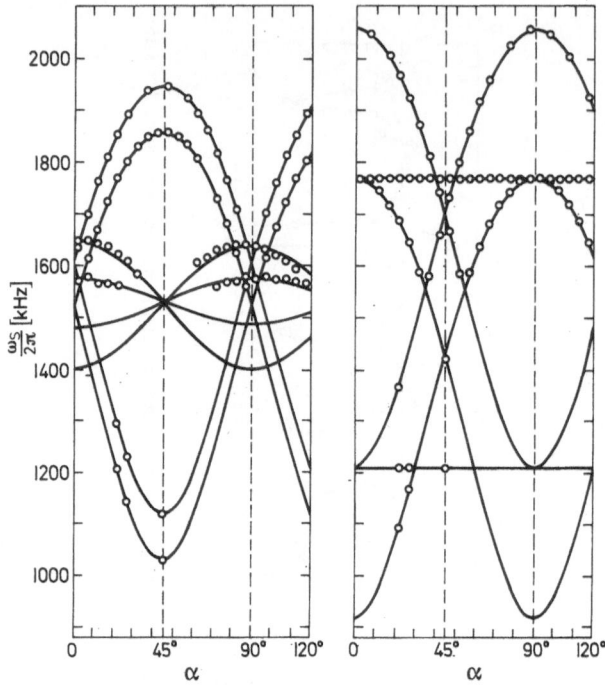

Fig. 16. Angular dependence of the DNMR lines in a single crystal of KF doped with NaF. α is the angle between the [100]-axis and $B_0(B_0 = 7.490\,\text{kG})$. Left: DNMR lines due to group A nuclei in Fig. 16a. Right: DNMR lines due to group B nuclei in Fig. 16a

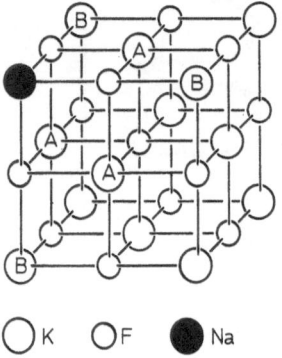

Fig. 16a. Model of the vicinity of a Na$^+$-impurity in KF

to determine the field gradients from second order quadrupole shifts of the central line, $m_S = 1/2 \leftrightarrow m_S = -1/2$. The asymmetry parameter η could not be determined in these experiments. But, for theoretical studies the asymmetry parameters are especially important since they do not depend on the Sternheimer antishielding factors in the crystal, which are unknown up to date.

Table IV.4 summarizes the experimental data about quadrupole splittings around impurities in alkali halides. These data yield information about the most interesting quantity, the lattice distortion around the impurity. Two fundamentally different starting points have been chosen to calculate the lattice distortion and the resulting field gradients on the basis of different models. The first method is based on a continuum model [57, 62–64]. The impurity ion is treated as a sphere in a hole, which results from removing an ion of the host lattice. The radii of sphere and hole are approximated by the radii of the two ions under consideration. The surrounding crystal is considered to be isotropic with the macroscopic elastic constants of the host crystal. (Sometimes even the impurity ion is assumed to have the macroscopic elastic constants [57–63].) In this way the tensor of the lattice distortion around the impurity can be calculated. This, in turn, yields the tensor of the field gradient, since both are related to each other by a fourth rank tensor. However, this continuum model should give realistic results for long range effects only. The model is inadequate for the calculation of the distortion next to the impurity. Therefore the agreement obtained by Andersson [57] for nearest neighbors of impurities in KBr and KI seems to be somewhat fortuitous.

Contrary to the continuum approximation Dick and Das [66–68] used a microscopic model for the calculation of the field gradients. They assumed that the impurity causes radial displacements of all the neighboring ions. The resulting change of the electric field can be described using fictitious charges of opposite signs, one of them is placed at the undistorted lattice site, the other one at the place of the ion in the distorted lattice. Furthermore, the ions are polarized. This causes a polarization dipole at the site of each ion. Dick [68] has shown that the ionic polarizabilities have to be modified compared to those of ions in the unperturbed crystal [69], because the outer electronic shell and the ionic core are displaced differently (shell model).

The most important contributions to the change in crystal energy arise from the nearest neighbor repulsion (Born-Mayer-potentials), the interaction between the electric dipols of the ions (displacement and polarization dipoles) and finally from the self energy of the ions because of the polarization. If this change of the energy is minimized the equilibrium positions, the dipolar moments, and the resulting field gradients can be calculated. The theoretical results are compared with the experimental results in columns 9–11 of Table IV.4. They take into account the enhancement of the field gradient because of the Sternheimer anti-

Table IV.4. Comparison of experimental and theoretical values of electric field gradients around impurities in alkali halides

Experimental							Theoretical				
Crystal	Impurity	Resonant nucleus ⟨Position⟩	e^2qQ/h (kHz)	η	$(e^2qQ/h)\sqrt{1-\frac{1}{3}\eta^2}$ (kHz)	eq (10^{14} esu)[a]	Method (Ref.)	eq (10^{14} esu)[a]	η	$(e^2qQ/h)\sqrt{1+\frac{1}{3}\eta^2}$ (kHz)	Ref.
1	2	3	4	5	6	7	8	9	10	11	12
NaF	Li	^{23}Na ⟨110⟩	544 ± 10	0.65 ± 0.05		0.751	KDR [24]	−1.18	0.705		[24]
	Li	^{23}Na ⟨200⟩	856 ± 10	0		1.18	KDR [24]	+1.32	0		[24]
NaF	K	^{23}Na ⟨110⟩	832 ± 10	0.45 ± 0.05		1.14	KDR [25]	+0.785	0.705		[25]
	K	^{23}Na ⟨200⟩	1520 ± 10	0		2.10	KDR [25]	−0.874	0		[25]
NaCl	K	^{23}Na ⟨110⟩			716	~1.0	NQDR [21]	−0.33	0.62	254	[67]
	K	^{23}Na ⟨112⟩			316	~0.43	NQDR [21]				
	K	^{35}Cl ⟨100⟩	2394			4.17	NQDR [21]	−2.46	0		[67]
	K	^{35}Cl ⟨111⟩	560			0.98	NQDR [21]	−0.87			
	Br⁻ [b]	^{23}Na ⟨100⟩	660 ± 20			0.91	[NMR [58] / NQDR [21]]	0.188	0		[67]
	Ag	^{23}Na ⟨200⟩	80 ± 1	0		0.111	KDR [18, 22]				
	Ag	^{35}Cl ⟨210⟩	44 ± 5	~0		0.076	KDR [22]				
KF	Na	^{39}K ⟨110⟩	916	0.607		1.81	KDR [23]	+1.42	0.784		[23]
	Na	^{39}K ⟨200⟩	1138	0		2.26	KDR [23]	−1.40	0		[23]
KBr	Na	^{79}Br ⟨100⟩	14030 ± 30	0		5.7	NMR [57]	+2.6			[67]
	Na	^{79}Br ⟨102⟩	1850 ± 50	0.45 ± 0.05		0.75	NMR [57]	+1.7	0.821		[67]
	Na	^{79}Br ⟨300⟩	890 ± 30	0		0.36	NMR [57]				
	Cl	^{79}Br ⟨110⟩	2440	0.9 ± 0.1		1.00	NMR [61]	−1.7	0.999		[67]
	Cl	^{79}Br ⟨200⟩	610	0		0.40	NMR [56]	−2.1			[67]

[a] To obtain the values in V/cm² multiply with 300, to obtain atomic units multiply with 0.154 · 10^{-15}.

[b] Tentative assignment.

Table IV. 5. Sternheimer antishielding factors $(1 - \gamma_\infty)$ and nuclear quadrupole moments used in Table IV.4

Ion	$1 - \gamma_\infty$	$Q(10^{-24}\ cm^2)$
^{23}Na	5.53	0.1
^{35}Cl	50.3	-0.079
^{39}K	13.8	0.07
^{79}Br	100	0.34

shielding factors, $(1 - \gamma_\infty)$. The values of $(1 - \gamma_\infty)$ and Q, used for this calculation are compiled in Table IV.5.

Again the agreement between the experimental and theoretical data is surprisingly good. However, DNMR experiments in NaF:Li and NaF:K [24, 25] gave results which are incompatible with the simple model of Dick and Das. Dick and Nelson [25] showed that the field gradients at equivalent lattice positions next to different impurity ions in the same host crystal should always have the same ratio, independent of ionic polarizabilities and antishielding factors. Also, the asymmetry parameters for all lattice positions should be independent of the particular impurity. Both predictions are not satisfied in NaF:Li and NaF:K, as shown in Table IV.4.

It is well known that the overlap between the electronic wave functions changes the charge density at the place of the nuclei [70–72]. The overlap depends severely on ionic displacements. Ikenberry and Das [73] calculated field gradients, which are produced at the sites of the neighboring ion nuclei because of the overlap. It can be shown that these contributions for nuclei in $\langle 100 \rangle$ and $\langle 200 \rangle$ positions exceed the contributions resulting from the displacements. For nuclei in $\langle 110 \rangle$ positions the overlap effects are relatively small. However, this expansion of the model did not improve the agreement between theoretical and experimental data considerably. Ikenberry and Das suggested that some covalent admixture in the chemical binding may transfer charge from the anion to the cation. This would change the antishielding factor effectively and improve the agreement. DNMR experiments can possibly clarify the type of binding in ionic crystals. However, there are not enough experimental data available until now to make a definite statement about the suggestion of Ikenberry and Das.

IV.4.C Second Order Quadrupole Effects

High field DNMR was especially useful in the investigation of quadrupole effects in samples with $I = \frac{1}{2}$. Alkali fluorides, for instance, were studied quite successfully using DNMR. The quadrupole interaction in systems

with $I = \frac{3}{2}$ could be studied using DNMR, if the interaction was extraordinary small (for instance NaCl:Ag) [18, 22]. (The observation of
significantly larger quadrupole interaction by use of DNMR in NaCl:Br
has been reported. However, it seems to be likely that the interpretation
of these lines was in error [75].) Until now, it is not definitely clear, why
S spins with larger quadrupole interactions could not be observed in
systems with $I > \frac{1}{2}$. We suppose that quadrupole interactions of the I spins
with the impurity are responsible for this failure. I spins with spin
$I > \frac{1}{2}$ possess a quadrupole moment and therefore are exposed to a
quadrupole interaction in the neighborhood of the impurity. The S
spins, however, have to be coupled to the unperturbed I spins. But
they interact only with their nearest neighbors, and spin diffusion [76]
must distribute the energy to the entire I spin system. The transition
frequencies of the neighboring nuclei depend on their distance from the
impurity in spin systems with $I > \frac{1}{2}$, since they are exposed to different
quadrupole interactions. This inhibits mutual flip flop processes and
may strongly reduce the spin diffusion. Spin diffusion can take place
only through the levels $|m_I\rangle = |\pm \frac{1}{2}\rangle$ if the quadrupole interaction is
small enough to make second order effects unimportant. If second order
quadrupole shifts are larger than the line width in the rotating frame
(which is smaller than that in the laboratory frame [77]), spin diffusion
may be severely reduced and the DNMR experiment cannot be successful.
A few experiments on a NaF crystal doped with KF support this hypothesis: The quadrupole shifts which could be easily observed by Nelson
and Ohlson [24] using the ^{19}F nuclei as I spins could not be detected
if the ^{23}Na nuclei where used as I spins even at the highest possible
sensitivity. We suppose that second order quadrupole effects inhibit
the spin diffusion so strongly that the disorder cannot be transferred
from the S spins to the I spin system.

IV.5. Application of DNMR to the Investigation of Phase Transitions

To complete this summary of experimental applications of DNMR, the
investigation of phase transitions should be mentioned. KH_2PO_4
undergoes a ferroelectric phase transition at a Curie temperature of
$128\,°K$. The proton resonance at this phase transition was studied by
several authors [78]. But since the protons don't have a quadrupole
moment these measurements do not yield information about the electric
field gradients in the crystal. Because of the small gyromagnetic ratio
of the ^{39}K nuclei their resonance signal cannot be observed with NMR
techniques. Tsutsumi and coworkers [29] therefore applied DNMR
methods using the ^{39}K nuclei as S spins and the protons as I spins. In

this way, they were able to measure the field gradients and asymmetry parameters at the site of the ^{39}K nuclei both in the parelectric and in the ferroelectric phase. They also calculated these quantities using a point ion lattice of the crystallographic structure as determined by neutron diffraction. They got good agreement for the parelectric phase. However, for the ferroelectric phase there is considerable disagreement between theory and experiment.

It is also interesting to study the spin lattice relaxation time of the ^{39}K nuclei [30] near the phase transition by means of DNMR. Without spin lattice relaxation the S spins would be spin locked after a single temperature exchange. However, the spin lattice relaxation of the S spins destroys this spin locking and permits an additional heat flow from the lattice to the I spin system via the S spins. Stehlik and Nordal measured the temperature dependence of T_1 of the ^{39}K nuclei in this way. They observed a strong increase of the inverse spin lattice relaxation time $1/T_1$ when approaching the transition temperature. Their results agree qualitatively with theoretical arguments which predicted a temperature dependence of $1/T_1 \propto (T - T_c)^{-0.5}$ near the transition temperature [79, 80]. Similar measurements of the spin lattice relaxation time could help to understand the ferroelectric phase transition.

V. Theory of Nuclear Double Resonance

V.1. The Concept of Spin Temperature

In the following chapter the theoretical methods shall be outlined which are necessary to understand DNMR quantitatively. We do not attempt to give a complete discussion, but rather to indicate the basic ideas as plainly as possible. Details are discussed in the extensive original literature [12, 13, 16, 18].

As a first point the concept of spin temperature shall be outlined briefly [81, 82]. Since the population probabilities p_n of states with energy E_n of a spin system in thermal equilibrium obey a Boltzmann distribution,

$$p_n = \eta_I^{-1} \exp - E_n/k\theta_I, \tag{V.1}$$

they can be characterized by a spin temperature θ_I. η_I is the partition function. If the Zeeman energy of a single spin is large compared to their mutual interaction, the concept of spin temperature is simply a convenient way to describe a spin system. If, however, the interaction between the spins is comparable with their Zeeman energy, individual eigenstates cannot be defined any more, but rather eigenstates of the entire spin sytem. This is proved by the following experiment [83]:

A LiF crystal, with spin lattice relaxation times in the order of minutes, is polarized in a high magnetic field to its equilibrium magnetization. Subsequently, it is removed from the magnetic field rapidly. If, after several seconds, it is put back into the magnetic field the equilibrium magnetization is restored immediately. Would it be possible to describe the system, even in zero field, by the eigenstates of individual spins, then the mutual dipolar interaction would cause an irreversible loss of order of the spin system. The reversibility of such an adiabatic demagnetization can be only explained by the assumption that the spin system is characterized by eigenstates of the entire system, and that their population probabilities obey a Boltzmann distribution with a spin temperature $\theta_I(B)$ for all magnetic fields.

This concept makes it possible to calculate the average values of physical quantities of the system in thermal equilibrium. This shall be indicated using a system of N_I nuclei of spin I and gyromagnetic ratio γ_I as an example. In an external magnetic field B_0 the Hamiltonian $\hat{\mathscr{H}}_I$ of this system consists of the Zeeman part

$$\hat{\mathscr{H}}_{ZI} = -\gamma_I \hbar B_0 \cdot \sum_p \hat{I}_p = -\gamma \hbar B_0 \cdot \hat{I}, \tag{V.2}$$

and the dipolar Hamiltonian

$$\hat{\mathscr{H}}_{dII} = \sum_{p,q \neq p} \hat{\mathscr{H}}_{dpq}$$

$$= \tfrac{1}{2}\mu_0(4\pi)^{-1} \sum_{p,q \neq p} \gamma_I^2 \hbar^2 [\hat{I}_p \cdot \hat{I}_q / r_{pq}^3 - 3(\hat{I}_p \cdot r_{pq})(\hat{I}_q \cdot r_{pq})/r_{pq}^5], \tag{V.3}$$

where r_{pq} is the vector from spin $\hat{I}_p$ to spin $\hat{I}_q$ and $\hat{I}$ is the vector sum of the spins of the individual nuclei. Whereas it is not feasible to solve the Schrödinger equation of such a system exactly, it is possible to calculate the ensemble averages in thermal equilibrium. For example, if one uses the relations

$$\hat{\mathscr{H}}_I |n\rangle = E_n |n\rangle, \tag{V.4}$$

and

$$\exp(-E_n/k\theta_I) |n\rangle = \exp(-\hat{\mathscr{H}}_I/k\theta_I) |n\rangle, \tag{V.5}$$

the average value of the energy is given by

$$E_I = \eta_I^{-1} \sum_n \langle n| \hat{\mathscr{H}}_I \exp(-\hat{\mathscr{H}}_I/k\theta_I) |n\rangle. \tag{V.6}$$

Thus the ensemble average is given by the sum of the diagonal elements (trace) of a matrix. Now it is well known that the trace is independent of the particular representation. One can therefore use any convenient representation, e.g. the one in which the basis functions are eigenfunctions of the individual nuclei. This procedure yields for the partition function

$$\eta_I = \sum_n \langle n| \, 1 \cdot \exp(-\hat{\mathscr{H}}_I/k\theta_I) \, |n\rangle = \text{Tr} \exp(-\hat{\mathscr{H}}_I/k\theta_I) . \qquad (V.7)$$

In high temperature approximation one can often expand the exponential in a power series and keep only the leading terms:

$$\eta_I = \text{Tr} \{1 - \hat{\mathscr{H}}_I/k\theta_I + \hat{\mathscr{H}}_I^2/2k^2\theta_I^2 - \cdots\}$$

$$= (2I + 1)^{N_I} + \tfrac{1}{2}k^{-2}\theta_I^{-2} \, \text{Tr}(\hat{\mathscr{H}}_I^2) . \qquad (V.8)$$

Here the fact was used that $\text{Tr}(\hat{\mathscr{H}}_I) = 0$, which can be readily verified for both $\hat{\mathscr{H}}_{zI}$ and $\hat{\mathscr{H}}_{dII}$, since the contributions of a pair of $|m_{I1} \ldots m_{Ij} \ldots m_{I_N}\rangle$ and $|m_{I1} \ldots -m_{Ij} \ldots m_{I_N}\rangle$ cancel each other mutually. As an example for the calculation of such traces, $\text{Tr}(\hat{\mathscr{H}}_I^2)$ shall be computed for zero external field. In this case

$$\text{Tr}(\hat{\mathscr{H}}_I^2) = \text{Tr}(\hat{\mathscr{H}}_{dII}^2) = \sum_{p,q \neq p} (\hat{\mathscr{H}}_{dpq})^2 . \qquad (V.9)$$

The evaluation of this expression is simple, if one considers the fact that in zero magnetic field there is no preferential axis. Thus, without loss of generality, the z-axis can be assumed to coincide with the direction of r_{ij}. Therefore

$$\hat{\mathscr{H}}_{dpq} = \tfrac{1}{2}(\mu_0/4\pi) \gamma_I^2 \hbar^2 \sum_{p,q \neq p} [\hat{I}_{xp}\hat{I}_{xq} + \hat{I}_{yp}\hat{I}_{yq} - 2\hat{I}_{zp}\hat{I}_{zq}] r_{pq}^{-3} . \qquad (V.10)$$

Using the relations [84]

$$\text{Tr}(\hat{I}_{xp}^2) = \tfrac{1}{3}I(I + 1)(2I + 1)^{N_I} , \qquad (V.11)$$

and

$$\text{Tr}(\hat{I}_{xp}^2 \hat{I}_{xq}^2) = \tfrac{1}{9}I^2(I + 1)^2 (2I + 1)^{N_I} , \qquad (V.12)$$

respectively, one obtains

$$\text{Tr}(\hat{\mathscr{H}}_{dpq}^2) = \tfrac{1}{3}(\mu_0/4\pi) \gamma_I^4 \hbar^4 I^2(I + 1)^2 (2I + 1)^{N_I} \sum_{p,q \neq p} r_{pq}^{-6}. \qquad (V.13)$$

This yields the final result

$$\text{Tr}(\hat{\mathscr{H}}_{dII}^2) = \tfrac{1}{3}N_I(\mu_0/4\pi) \gamma_I^4 \hbar^4 I^2(I + 1)^2 (2I + 1)^{N_I} \sum_{p,q \neq p} r_{pq}^{-6} . \qquad (V.14)$$

A quantity which is often used is the local field [15, 36, 85, 86] defined by

$$B_{\text{loc}}^2 = \frac{\mu_0}{kC_I(2I + 1)^{N_I}} \, \text{Tr}(\hat{\mathscr{H}}_{dII}^2) , \qquad (V.15)$$

where C_I is the Curie constant of the ensemble:

$$C_I = N_I \gamma_I^2 \hbar^2 I(I+1) \mu_0/3k . \qquad (V.16)$$

Combining Eqs. (V.14) and (V.15) one obtains

$$B_{\text{loc}}^2 = (\mu_0/4\pi) \gamma_I^2 \hbar^2 I(I+1) \sum_{p,q \neq p} r_{pq}^{-6} . \qquad (V.17)$$

Equation (V.15) is the exact definition for the local field, which has been introduced qualitatively already in Section II.3. Similarly, traces of other quantities can be calculated in a convenient representation. The partition function, the energy, and the magnetization of a system of N_I spins in a magnetic field B_0 are, for instance, given by Eqs. (V.18)–(V.20), respectively:

$$\eta_I = \text{Tr}(1 \cdot \hat{\varrho})$$

$$= (2I+1)^{N_I} + \frac{C_I}{2k\theta_I \mu_0} (B_0^2 + B_{\text{loc}}^2) \approx (2I+1)^{N_I} , \qquad (V.18)$$

$$E_I = -(C_I/\mu_0)(B_0^2 + B_{\text{loc}}^2)/\theta_I , \qquad (V.19)$$

$$M_I = (C_I/\mu_0)(B_0/\theta_I) . \qquad (V.20)$$

The operator

$$\hat{\varrho} = \eta_I^{-1} \exp(-\hat{\mathscr{H}}_I/k\theta_I) , \qquad (V.21)$$

which was introduced in Eq. (V.5) is the so called density operator of the ensemble. It is the operator analogon of the Boltzmann distribution, Eq. (V.1). It containes all the information necessary to describe a statistical ensemble of identical particals in thermal equilibrium. $\hat{\varrho}$ can be represented by the so called density matrix with the matrix elements $\varrho_{mm'} = \langle m| \hat{\varrho} |m' \rangle$.

The concept of the density matrix is discussed in great detail in the books of Abragam, Slichter and Goldman, expecially with regard to its application to spin systems. Therefore only the most important properties shall be summarized:

a) The ensemble average of any physical quantity $\hat{O}$ is given by

$$\langle O \rangle = \text{Tr}(\hat{O} \cdot \hat{\varrho}) . \qquad (V.22)$$

Thus, if the density matrix of the system is known, it is possible to calculate all the physical properties of the ensemble. Since the trace is independent of the representation, this relation is most important for all practical purposes.

b) The time dependence of $\hat{\varrho}$ is given by

$$d\hat{\varrho}/dt = (i/\hbar)[\hat{\mathscr{H}}, \hat{\varrho}] = (i/\hbar)[\hat{\mathscr{H}}\hat{\varrho} - \hat{\varrho}\hat{\mathscr{H}}] . \qquad (V.23)$$

(It might be worthwhile to mention that the sign of this relation is opposite to the Heisenberg operator equation for the time derivative of an observable.) In the event that $\hat{\mathscr{H}}$ is time independent a formal solution of Eq. (V.23) is readily obtained:

$$\hat{\varrho}(t) = e^{-i\hat{\mathscr{H}}t/\hbar} \hat{\varrho}(0) e^{i\hat{\mathscr{H}}t/\hbar}. \tag{V.24}$$

If $\hat{\varrho}$ depends only on $\hat{\mathscr{H}}$, it commutes with $\hat{\mathscr{H}}$ and, according to Eq. (V.23), $\hat{\varrho}$ is time independent, $d\hat{\varrho}/dt = 0$.

c) For a canonical ensemble, whose energy levels are occupied according to a Boltzmann distribution, the density matrix is given by Eq. (V.21). In thermal equilibrium all the off-diagonal elements of the density matrix vanish. This is a consequence of the hypothesis of random phases, which postulates that the phases of the wave functions of the individual particles are statistically independent. In the event that there are nonvanishing off-diagonal elements, for instance immediately after a 90° pulse, the system cannot be described by a spin temperature. However, if the off-diagonal elements vanish at any time then, according to Eq. (V.23), they will vanish for all times, and the system can be described by a spin temperature.

The property c) is particularly important for the theory of nuclear double resonance. It is the basis for the concept of spin temperature in the rotating frame. In the presence of a strong rotating rf field whose frequency roughly satisfies the resonance condition, $\omega_I \approx \gamma_I B_0$, the Hamiltonian contains in addition to the static Zeeman term, Eq. (V.2), and the dipolar interaction operator, Eq. (V.3), a time dependent contribution

$$\hat{\mathscr{H}}_{rf} = -\gamma_I B_{1I}\hbar(\cos \omega_I t \cdot \hat{I}_x - \sin \omega_I t \cdot \hat{I}_y), \tag{V.25}$$

In the rotating frame the interaction with the external fields B_0 and B_1 is time independent. In fact, in this frame the dipolar interaction includes time dependent (non-secular) parts, but their frequency $\pm \omega_I$ and $\pm 2\omega_I$ are very large compared to the resonance frequency in the rotating frame

$$\omega_{eff} = \gamma_I B_{eff} \approx \gamma_I B_{1I}. \tag{V.26}$$

Therefore these parts can be neglected in most cases. The basic assumption in Redfields concept [15] is that the population probabilities in the rotating frame of the states of the system under the influence of the quasistationary transformed Hamiltonian obey the same laws as the population probabilities in the laboratory frame under the influence of a time independent Hamiltonian. This means that the ensemble approaches its most probable state and in thermal equilibrium has a canonical distribution. Thus it can be characterized by a density matrix

$$\hat{\varrho}' = \eta^{-1} e^{-\hat{\mathscr{H}}'/k\theta'}, \tag{V.27}$$

where $\hat{\varrho}'$, $\hat{\mathscr{H}}'$ and $\hat{\theta}'$ are the corresponding quantities transformed into the rotating frame. Several experiments [36, 87, 88], last not least DNMR, have proved this hypothesis.

V.2. Transformation into the Rotating Frame

In order to employ the methods of thermodynamics the spin Hamiltonian has to be transformed into the rotating frame. Consider a solid containing two nuclear spin species I and S in a strong external magnetic field B_0, which is exposed to the two circular polarized rf fields $B_{1I}(x \cdot \cos \omega_I t - y \cdot \sin \omega_I t)$ and $B_{1S}(x \cdot \cos \omega_S t - y \cdot \sin \omega_S t)$, respectively. ω_I and ω_S are frequencies close to the resonance frequencies $\gamma_I B_0$, and $\gamma_S B_0$, respectively. The total Hamiltonian for the nuclear system is

$$\hat{\mathscr{H}} = \hat{\mathscr{H}}_z + \hat{\mathscr{H}}_{rf} + \hat{\mathscr{H}}_d, \tag{V.28}$$

where

$$\hat{\mathscr{H}}_z = -\hbar B_0 \left[\gamma_I \sum_{p=1}^{N_I} \hat{I}_{zp} + \gamma_S \sum_{j=1}^{N_S} \hat{S}_{zj} \right]. \tag{V.29}$$

$\hat{\mathscr{H}}_d$ is the sum of the interaction of the I spins themselves, $\hat{\mathscr{H}}_{dII}$, the interaction between the I and the S spins, $\hat{\mathscr{H}}_{dIS}$, and the interaction between the S spins themselves, $\hat{\mathscr{H}}_{dSS}$. For the following discussion we may write the dipolar Hamiltonian according to van Vleck [89] in a form particularly convenient for computing matrix elements

$$\hat{\mathscr{H}}_{dII} = \tfrac{1}{2}(\mu_0/4\pi)\gamma_I^2\hbar^2 \sum_{p,q \neq p} r_{pq}^{-3}(A_{pq} + B_{pq} + C_{pq} + D_{pq} + E_{pq} + F_{pq}),$$

$$\tag{V.30}$$

where

$$A_{pq} = \hat{I}_{zp}\hat{I}_{zq}(1 - 3\cos^2\vartheta_{pq}), \tag{V.31a}$$

$$B_{pq} = -\tfrac{1}{4}(\hat{I}_p^+\hat{I}_q^- + \hat{I}_p^-\hat{I}_q^+)(1 - 3\cos^2\vartheta_{pq}), \tag{V.31b}$$

$$C_{pq} = -\tfrac{3}{2}(\hat{I}_p^+\hat{I}_{zq} + \hat{I}_{zp}\hat{I}_q^+)\sin\vartheta_{pq}\cos\vartheta_{pq}\,e^{-i\varphi_{pq}}, \tag{V.31c}$$

$$D_{pq} = -\tfrac{3}{2}(\hat{I}_p^-\hat{I}_{zq} + \hat{I}_{zp}\hat{I}_q^-)\sin\vartheta_{pq}\cos\vartheta_{pq}\,e^{i\varphi_{pq}}, \tag{V.31d}$$

$$E_{pq} = -\tfrac{3}{4}\hat{I}_p^+\hat{I}_q^+\sin^2\vartheta_{pq}\,e^{-2i\varphi_{pq}}, \tag{V.31e}$$

$$F_{pq} = -\tfrac{3}{4}\hat{I}_p^-\hat{I}_q^-\sin^2\vartheta_{pq}\,e^{2i\varphi_{pq}}. \tag{V.31f}$$

r_{pq} is the vector connecting the spins p and q, ϑ_{pq} the angle between r_{pq} and the z-axis, and φ_{pq} the angle between the projection of r_{pq} into the xy-plane and the x-axis. $\hat{I}_p^{\pm}$ are the raising or lowering operators $(\hat{I}_{xp} \pm i\hat{I}_{yp})$. $\mathscr{H}_{dSS}$ is defined in a similar way, by substituting S spin operators in place of I spin operators.

$\mathscr{H}_{dIS}$ is given by

$$\mathscr{H}_{dIS} = (\mu_0/4\pi)\,\gamma_I\gamma_S\hbar^2 \sum_{p=1}^{N_I} \sum_{j=1}^{N_S} r_{pj}^{-3}\,(A_{pj} + B_{pj} + C_{pj} + D_{pj} + E_{pj} + F_{pj})\,.$$

$$(V.32)$$

The coefficients A_{pj}, etc., are the same as given in Eqs. (V.31) with the substitution $\hat{S}_j$ for $\hat{I}_q$. In addition we define the vectorsums

$$\hat{I} = \sum_{p=1}^{N_I} \hat{I}_p\,,$$

$$(V.33a)$$

and

$$\hat{S} = \sum_{j=1}^{N_S} \hat{S}_j\,.$$

$$(V.33b)$$

If the difference between ω_S and ω_I is large enough, the influence of B_{1S} on the I spins and of B_{1I} on the S spins can be neglected and $\mathscr{H}_{rf}$ can be approximated by

$$\begin{aligned}
\mathscr{H}_{rf} &\approx -\gamma_I\hbar B_{1I}(\hat{I}_x \cos \omega_I t - \hat{I}_y \sin \omega_I t) \\
&\quad - \gamma_S\hbar B_{1S}(\hat{S}_x \cos \omega_s t - \hat{S}_y \sin \omega_s t) \\
&= -\gamma_I\hbar B_{1I}\,e^{i\omega_I t \cdot \hat{I}_z}\,\hat{I}_x\,e^{-i\omega_I t \cdot \hat{I}_z} \\
&\quad - \gamma_S\hbar B_{1S}\,e^{i\omega_s t \cdot \hat{S}_z}\,\hat{S}_x\,e^{-i\omega_s t \cdot \hat{S}_z}\,.
\end{aligned}$$

$$(V.34)$$

In the second part of Eq. (V.34) the well known operator relation [90]

$$\hat{I}_x \cos \omega_I t - \hat{I}_y \sin \omega_I t = e^{i\omega_I t \cdot \hat{I}_z}\,\hat{I}_x\,e^{-i\omega_I t \cdot \hat{I}_z}\,,$$

$$(V.35)$$

has been employed.

The substitution of the wave function

$$\psi' = \hat{T}\psi = e^{-i\omega_I t \cdot \hat{I}_z}\,e^{-i\omega_s t \cdot \hat{S}_z}\,\psi\,,$$

$$(V.36)$$

into the Schrödinger equation

$$-(\hbar/i)\,(\partial\psi/\partial t) = \mathscr{H}\psi\,,$$

$$(V.37)$$

is equivalent to the classical rotating frame transformation [91]. This substitution gives a new Schrödinger equation:

$$-(\hbar/i)\,(\partial\psi'/\partial t) = \mathscr{H}'\psi'\,,$$

$$(V.38)$$

where

$$\hat{\mathscr{H}}' = \hat{T}\,\hat{\mathscr{H}}\,\hat{T}^{-1} - i\hat{T}\,\partial(\hat{T}^{-1})/\partial t$$

$$= -\gamma_I \hbar[(B_0 - \omega_I/\gamma_I)\,\hat{I}_z + B_{1I}\hat{I}_{x'}]$$

$$\quad - \gamma_S \hbar[(B_0 - \omega_S/\gamma_S)\,\hat{S}_z + B_{1S}\hat{S}_{x'}]$$ (V.39)

$$\quad + \hat{\mathscr{H}}_{dII}^0 + \hat{\mathscr{H}}_{dIS}^0 + \hat{\mathscr{H}}_{dSS}^0$$

+ time dependent parts which result from the dipolar coupling.

By applying the transformation Eq. (V.36) the Hamiltonian is transformed into a doubly rotating frame. The I and S spins are viewed in two different coordinate systems which rotate around the common z-axis at frequencies ω_I and ω_S, respectively. The two spin ensembles are coupled through the nonsecular part of their mutual dipolar interaction $\hat{\mathscr{H}}_{dIS}^0$. The two first terms of $\hat{\mathscr{H}}'$ will be denoted as $\hat{\mathscr{H}}_{ZI}'$ and $\hat{\mathscr{H}}_{ZS}'$. $\hat{\mathscr{H}}_{dII}^0$, $\hat{\mathscr{H}}_{dIS}^0$ and $\hat{\mathscr{H}}_{dSS}^0$ are the nonsecular parts of the dipolar coupling:

$$\hat{\mathscr{H}}_{dII}^0 = \tfrac{1}{4}(\mu_0/4\pi)\,\gamma_I^2\hbar^2 \sum_{p,q\neq p} r_{pq}^{-3}(1 - 3\cos^3 \vartheta_{pq})\,(3\hat{I}_{zp}\hat{I}_{zq} - \hat{I}_p\cdot\hat{I}_q),$$ (V.40)

$$\hat{\mathscr{H}}_{dIS}^0 = (\mu_0/4\pi)\,\gamma_I\gamma_S\hbar^2 \sum_{p=1}^{N_I}\sum_{j=1}^{N_S} r_{pj}^{-3}(1 - 3\cos^2 \vartheta_{pj})\,\hat{I}_{zp}\hat{S}_{zj},$$ (V.41)

and similarly for $\hat{\mathscr{H}}_{dSS}^0$.

The time dependent terms in Eq. (V.39) oscillate at frequencies $(\omega_I - \omega_S)$, $(\omega_I + \omega_S)$, ω_I, $2\omega_I$, ω_S and $2\omega_S$, which all are large compared to the resonance frequencies in the rotating frame,

$$\omega_{1I} = \gamma_I \sqrt{(B_0 - \omega_I/\gamma_I)^2 + B_{1I}^2} \approx \gamma_I B_{1I},$$

and   (V.42)

$$\omega_{1S} = \gamma_S \sqrt{(B_0 - \omega_S/\gamma_S)^2 + B_{1S}^2} \approx \gamma_S B_{1S}.$$

Therefore they are only small perturbations, which usually can be neglected. Thus $\hat{\mathscr{H}}'$ is essentially time independent, and its individual parts can be regarded as different energy reservoirs. According to Redfield's basic hypothesis these reservoirs will eventually reach a state of internal equilibrium, which can be characterized by a common spin temperature in the rotating frame and therefore by a density matrix

$$\varrho' = e^{-\hat{\mathscr{H}}'/k\theta'}/\mathrm{Tr}(e^{-\hat{\mathscr{H}}'/k\theta'}).$$ (V.43)

Using this density matrix the average values of the operators in the rotating frame can be calculated. One finds, for instance,

$$\langle E' \rangle = \mathrm{Tr}(\hat{\varrho}' \cdot \hat{\mathscr{H}}') = - \frac{C_I}{\mu_0} \frac{(B_{\mathrm{eff},I}^2 + B_{\mathrm{loc}}'^2) + C_S B_{\mathrm{eff},S}^2}{\theta'}, \tag{V.44}$$

$$M_I' = (C_I/\mu_0)(B_{\mathrm{eff},I}/\theta'), \tag{V.45}$$

as the equivalent to Eqs. (V.19) and (V.20), respectively. C_I and C_S are the Curie constants for the I and S spin systems, respectively. B_{loc}' is the local field in the rotating frame defined by

$$C_I B_{\mathrm{loc}}'^2/\theta' = \mathrm{Tr}\{\hat{\varrho}' \cdot (\hat{\mathscr{H}}_{dII}^0 + \hat{\mathscr{H}}_{dIS}^0 + \hat{\mathscr{H}}_{dSS}^0)\}. \tag{V.46}$$

It is related to the second moments of the NMR lines [86, 89] by

$$\gamma_I^2 B_{\mathrm{loc}}'^2 = \frac{1}{3} \langle \Delta^2 \omega_{II} \rangle + \langle \Delta^2 \omega_{IS} \rangle + \frac{\gamma_I^2 N_S S(S+1)}{3\gamma_S^2 N_I I(I+1)} \langle \Delta^2 \omega_{SS} \rangle, \tag{V.47}$$

where $\langle \Delta^2 \omega_{AB} \rangle$ is the B spin contribution to the second moment of the A spins. Note, that B_{loc}' is different from the local field in the laboratory frame, B_{loc}. The latter one is given by the trace of the total dipolar interaction operator, whereas only secular parts are included here.

The energy reservoir can reach a common temperature only if they can exchange energy. The efficiency of this exchange depends on two factors: First, on the matching of the energy levels, and second, on the tightness of the coupling. Two reservoirs can exchange energy only if their respective Hamiltonians do not commute [3]. For instance the I spin Zeeman reservoir cannot exchange energy with the S spin Zeeman reservoir directly, since $[\hat{\mathscr{H}}_{ZI}', \hat{\mathscr{H}}_{ZS}'] = 0$. If $B_{1I} \neq 0$ and $B_{1S} = 0$, $\hat{\mathscr{H}}_{ZI}'$ and $\hat{\mathscr{H}}_d'$ do not commute, and are coupled to a common energy reservoir. But $\hat{\mathscr{H}}_{ZS}'$ is still decoupled from this reservoir and the I and S spin systems can be regarded separately with respective heat capacities:

$$C_{B,I} = (\partial \langle E_I' \rangle/\partial \theta_I')_B = C_I(B_{\mathrm{eff},I}^2 + B_{\mathrm{loc}}'^2)/\mu_0 \theta_I'^2, \tag{V.48}$$

and

$$C_{B,S} = (\partial \langle E_S' \rangle/\partial \theta_S')_B = C_S B_{\mathrm{eff},S}^2/\mu_0 \theta_S'^2. \tag{V.49}$$

If $B_{1S} \neq 0$, $\hat{\mathscr{H}}_{ZS}'$ and $\hat{\mathscr{H}}_{dIS}^0$ do not commute any more and can exchange energy. This means that now the S spin system can exchange energy with the I spin reservoir.

---

[3] If the total system consists of two systems A and B with Hamiltonians $\hat{\mathscr{H}}_A$ and $\hat{\mathscr{H}}_B$ which commute with each other, each of them commutes with the total Hamiltonian and therefore is time-independent. Thus the systems cannot exchange energy.

V.3. Maximum Sensitivity

In this section the maximum sensitivity of the DNMR experiment shall be estimated regardless of the dynamics. Spin lattice relaxation shall be neglected.

Consider first the spin locking of the I spin system accomplished by a 90° pulse with a subsequent 90° phase shift: At $t = 0$ the system is in thermal equilibrium with the lattice in a strong magnetic field $B_0 \gg B_{loc}$. Its density matrix is

$$\hat{\varrho}_L = \hat{\varrho}' \approx \eta_I^{-1} \exp(\gamma_I \hbar B_0 \hat{I}_z / k\theta_l), \tag{V.50}$$

where θ_l is the lattice temperature. Now a 90° pulse is applied to the sample, i.e. for $0 \leq t \leq \pi/2\gamma_I B_{1I}$ the Hamiltonian of the system in the rotating frame is

$$\hat{\mathscr{H}}'_{ZI} = -\gamma_I \hbar B_{1I} \hat{I}_{x'}, \tag{V.51}$$

(assuming that $B_{1I} \gg B'_{loc}$). The equation of motion of the density matrix is then

$$d\hat{\varrho}'/dt = (i/\hbar) [\hat{\mathscr{H}}'_{ZI}, \hat{\varrho}'], \tag{V.52}$$

which can be integrated to yield

$$\hat{\varrho}'(t = \pi/2\gamma_I B_{1I}) = \eta_I^{-1} e^{-\frac{1}{2}i\pi \hat{I}_{x'}} e^{\gamma_I \hbar B_0 \hat{I}_z / k\theta_l} e^{\frac{1}{2}i\pi \hat{I}_{x'}}$$
$$= \eta_I^{-1} \exp(\gamma_I \hbar B_0 \hat{I}_{y'} / k\theta_l). \tag{V.53}$$

Subsequently, the only nonvanishing term of the Hamiltonian in the rotating frame is the dipolar interaction, which until now was neglected because of the assumption $B_{1I} \gg B_{loc}$. It causes the free induction decay of the transverse magnetization. If, however, at the end of the 90° pulse the phase of B_{1I} is shifted by 90°, so that $\hat{\mathscr{H}}'_{ZI} = -\gamma_I \hbar B_{1I} \hat{I}_{y'}$ the equation of motion is

$$d\hat{\varrho}'/dt = (i/\hbar) [-\gamma_I \hbar B_{1I} \hat{I}_{y'}, \quad \eta_I^{-1} \exp(\gamma_I \hbar B_0 \hat{I}_{y'} / k\theta_l)] = 0. \tag{V.54}$$

Thus $\hat{\varrho}'$ now is time independent with diagonal elements obeying a Boltzmann distribution. Using $\hat{\varrho}'$ the average value of the energy can be calculated:

$$\langle E'_I \rangle = \mathrm{Tr}(\hat{\mathscr{H}}'_{ZI} \cdot \hat{\varrho}') = -\tfrac{1}{3} N_I \gamma_I^2 \hbar^2 I(I+1) B_{1I} B_0 / k\theta_l = -C_I B_{1I}^2 / \mu_0 \theta'_{I0}. \tag{V.55}$$

Equation (V.55) defines the spin temperature in the rotating frame,

$$\theta'_{I0} = B_{1I} \theta_l / B_0. \tag{V.56}$$

This definition illustrates the equivalence between the rotating frame and the laboratory frame. Equation (V.54) explains the phenomenon of spin locking. If now B_{1I} is reduced adiabatically from amplitudes large compared to B'_{loc} to values comparable with B'_{loc} or less, this yields an adiabatic demagnetization in the rotating frame equivalent to the adiabatic demagnetization in the laboratory frame. The corresponding values of the energy and the spin temperature in the rotating frame are then:

$$\langle E'_I \rangle = - C_I(B_{1I}^2 + B'^2_{\text{loc}})/\mu_0 \theta'_{I0}, \tag{V.57}$$

and

$$\theta'_{I0} = \sqrt{B_{1I}^2 + B'^2_{\text{loc}}}\, \theta_I/B_0. \tag{V.58}$$

The disorder of the S spin system can be produced, for instance, by a sudden switching of B_{1S} (Section III.2). Assume that its frequency ω_S deviates from $\gamma_S B_0$ by the amount, $\Delta\omega_S = (\omega_S - \gamma_S B_0)$, and that the system originally was polarized to its equilibrium magnetization, $M_{0S} = C_S B_0/\mu_0 \theta_I$.

If B_{1S} is switched on suddenly the state of the S system does not change. Thus the density matrix after the switching process is the same as before. The Zeeman energy of the S system immediately after switching on B_{1S} is therefore

$$\langle E'_S \rangle = - M_0 \cdot B_{\text{eff},S} = -(C_S B_0/\mu_0 \theta_I)\, \Delta\omega_S/\gamma_S. \tag{V.59}$$

If $\Delta\omega_S = 0$, the energy vanishes after the sudden switching. Since both spin systems are coupled via $\mathcal{H}^0_{dIS}$ they will eventually reach a common temperature. Conservation of the total energy requires that

$$\frac{C_I(B_{1I}^2 + B'^2_{\text{loc}})}{\mu_0 \theta'_{I0}} = \frac{C_I(B_{1I}^2 + B'^2_{\text{loc}}) + C_S B_{1S}^2}{\mu_0 \theta'_{I1}}, \tag{V.60}$$

where θ'_{I1} is the common final temperature. Equation (V.60) yields

$$\theta'_{I0}/\theta'_{I1} = (1 + \varepsilon)^{-1}, \tag{V.61}$$

with

$$\varepsilon = (C_S B_{1S}^2/C_I)(B_{1I}^2 + B'^2_{\text{loc}})^{-1} \tag{V.62}$$

ε is the ratio of the respective specific heats at the final temperature. It is easy to show that switching off B_{1S} does not change the energy of the system. Thus the spin temperature does not change. This argument can be repeated for the following on-off-cycles provided that the spin lattice relaxation is negligible. After n cycles the ratio of the temperatures is then

$$\theta'_{I0}/\theta'_{In} = (1 + \varepsilon)^{-n} \approx e^{-n\varepsilon}. \tag{V.63}$$

The second relation in Eq. (V.63) holds, since in the experiments n is large and $\varepsilon \ll 1$.

The analyis of the experiment, in which the S spin system is heated by periodic $180°$ phase shifts, is essentially the same. However, since the energy of the S spin system changes its sign during the phase shift, this method is twice as efficient. Therefore the rise of the temperature of the I spin system after n phase shifts is given by

$$\theta'_{I0}/\theta'_{In} = e^{-2n\varepsilon}. \tag{V.64}$$

Equation (V.64) has been confirmed experimentally by several authors [13, 16–18]. CaF_2($^{43}Ca = S$ spins, $^{19}F = I$ spins) has been studied most extensively [16]. If the I spins are totally demagnetized in the rotating frame, they are exposed only to the local field, which for $\langle 111 \rangle$ parallel to B_0 amounts to $B'_{loc} = 0.86$ G. For $B_{1S} = 12$ G the double resonance condition ($\gamma_S B_{1S} \approx \gamma_I B'_{loc}$) is satisfied. In this case the I and S spin systems reach the common temperature after approximately 3 msec (Section V.4). After a single temperature exchange the I spin temperature is increased by about 1.4% ($\varepsilon \approx 1.4 \cdot 10^{-2}, \gamma_S = 1800$ sec^{-1} G$^{-1}, \gamma_I = 25\,170$ sec^{-1} G^{-1}, $N_S/N_I = 6.5 \cdot 10^{-4}$, $S = 7/2$, $I = 1/2$). The corresponding decrease of the I spin signal can hardly be detected. Since, however, the life time of the adiabatic demagnetized state amounts to about 4 sec the exchange process can be repeated more than 1000 times and in principal ^{43}Ca nuclei at a concentration of $10^{-4}\%$ could be detected.

V.4. Dynamics of the Temperature Exchange

In the previous section a complete temperature exchange was assumed during each heating cycle. The thermal conductivity between the S and I spin system determines the time required and therefore the sensitivity of DNMR. The temperature exchange in the DNMR process can be described as cross relaxation in the doubly rotating frame, analogous to the heat exchange between different spin systems in the laboratory frame, which has been studied by numerous authors [92–96]. The dynamics of this process is extraordinary complicated and explained only partially, although it was studied with great effort [12, 13, 16, 18]. The most intuitive model for the calculation of the exchange time T_{IS} has been suggested by Lurie and Slichter [13], in close relation to the work of Schumacher and Bloembergen et al. [93, 94].

The following discussion is restricted to the case of exact resonance, $\omega_S = \gamma_S B_0$. Two basic assumptions will be made:

a) The coupling between the I and S spin system is assumed to be weak enough that $\hat{\mathscr{H}}^0_{dIS}$ is a small perturbation to $\hat{\mathscr{H}}'_I$ and $\hat{\mathscr{H}}'_S$, respectively. This assumption is a serious restriction. However, until now only first order perturbation theory could be applied successfully for the calculation of T_{IS}.

b) Both, the I and S spin systems can be described by spin temperatures, θ'_I and θ'_S, respectively.

In the event that the first assumption is satisfied ($T_{IS} \gg T_{2I}$) there exists a uniform I spin temperature. (Deviations from a uniform spin temperature will be discussed in connection with the influence of spin diffusion in part VI.) Because of the low S spin concentration their mutual interaction is only weak. In spite of this, the concept of an S spin temperature can be justified, since the S spins have a uniform temperature at the beginning of the exchange process. Furthermore, the transition matrix elements of the shape $|\langle m|\hat{S}_x|m'\rangle|^2$ conserve an existing Boltzmann distribution.

Consider first the temperature exchange after a single heating of the S spin system. The cross relaxation can be studied most easily, if the energy of the S system is regarded as the energy of a macroscopic magnetization M_S in a field B_{1S}. The relaxation of the S spin system is described by the rate equation

$$\mathrm{d}M_S(\theta'_S)/\mathrm{d}t = -R_{SI}(M_S(\theta'_S) - M_S(\theta'_I)),\tag{V.65}$$

where R_{SI} is the rate constant for the temperature exchange, which shall be estimated. Application of Curie's law transforms Eq. (V.65) into

$$\mathrm{d}\beta_S/\mathrm{d}t = -R_{SI}(\beta_S - \beta_I),\tag{V.66}$$

where $\beta_S = 1/k\theta'_S$. A similar equation can be given for the I spin temperature

$$\mathrm{d}\beta_I/\mathrm{d}t = -R_{IS}(\beta_I - \beta_S).\tag{V.67}$$

Because of the conservation of energy R_{IS} and R_{SI} are related to each other by

$$R_{IS} = \varepsilon R_{SI}.\tag{V.68}$$

If Eq. (V.66) is subtracted from Eq. (V.67) one obtains a differential equation for the difference of the inverse spin temperatures

$$\begin{aligned}\mathrm{d}(\beta_I - \beta_S)/\mathrm{d}t &= -(R_{IS} + R_{SI})(\beta_I - \beta_S)\\ &= -(1/T_{IS})(\beta_I - \beta_S).\end{aligned}\tag{V.69}$$

Thus the equilibrium between both systems is approached with a uniform time constant T_{IS} [93]. In the following, this time constant shall be estimated by means of first order perturbation theory. For this purpose we consider the energy of the I spin system in high temperature approximation

$$\langle E_I' \rangle = \mathrm{Tr}(\hat{\varrho}' \cdot \hat{\mathscr{H}}_I') \approx -\eta_I^{-1} \beta_I \, \mathrm{Tr}(\hat{\mathscr{H}}_I'^2), \tag{V.70}$$

where $\eta_I = (2I+1)^{N_I}$ is the partition function of the I spin system. ($\hat{\mathscr{H}}_I'$ is defined by Eq. (V.83).) Differentiation of Eq. (V.70) yields

$$\mathrm{d}\langle E_I' \rangle/\mathrm{d}t = (\mathrm{d}\langle E_I' \rangle/\mathrm{d}\beta_I)\,(\mathrm{d}\beta_I/\mathrm{d}t) = -\eta_I^{-1}\,\mathrm{Tr}(\hat{\mathscr{H}}_I'^2)\,\mathrm{d}\beta_I/\mathrm{d}t, \tag{V.71}$$

on the other hand

$$\langle E_I' \rangle = \sum_n p_n E_n'. \tag{V.72}$$

Thus

$$\mathrm{d}\langle E_I' \rangle/\mathrm{d}t = \sum_n E_n'\,\mathrm{d}p_n/\mathrm{d}t, \tag{V.73}$$

where p_n is the population probability of the eigenstate $|n\rangle$ of energy E_n'. If the system can be characterized by a temperature, the population probabilities obey a Boltzmann distribution, which in high temperature approximation can be written as

$$p_n = p_m[1 - (E_n' - E_m')\,\beta_I]. \tag{V.74}$$

Similarly the population probabilities of the states $|u\rangle$ of the S spin system with energies E_u' are given as

$$q_u = q_v[1 - (E_u' - E_v')\,\beta_S]. \tag{V.75}$$

The occupation numbers are changed only by mutual flipping processes between the I and S systems. We define $W_{mu,nv}$ as the probability for a simultaneous transition of the I system from the state $|m\rangle$ to the state $|n\rangle$ and the S system from $|u\rangle$ to $|v\rangle$, respectively. Then

$$\mathrm{d}p_n/\mathrm{d}t = \sum_{mu,nv} W_{mu,nv}(p_m q_u - p_n q_v), \tag{V.76}$$

(since $W_{nv,mu} = W_{mu,nv}$). Energy conservation requires that $(E_u' - E_v') = (E_n' - E_m')$. Thus Eqs. (V.74–V.76) can be combined to give

$$\mathrm{d}p_n/\mathrm{d}t = \sum_{mu,nv} W_{mu,nv}(E_n' - E_m')\,p_m q_v(\beta_I - \beta_S). \tag{V.77}$$

In order to treat the labels m and n more symmetrically one can rewrite Eq. (V.73):

$$\mathrm{d}\langle E_I' \rangle/\mathrm{d}t = \tfrac{1}{2}\left[\sum_n E_n'\,\mathrm{d}p_n/\mathrm{d}t + \sum_m E_m'\,\mathrm{d}p_m/\mathrm{d}t\right]. \tag{V.78}$$

If we now substitute Eq. (V.77) and use the approximations

$$p_n \approx 1/\eta_I, \quad \text{and} \quad q_u \approx 1/\eta_S, \tag{V.79}$$

we get

$$d\langle E_I' \rangle / dt = (2\eta_I \eta_S)^{-1} \sum_{mu,nv} W_{mu,nv} (E_n' - E_m')^2 (\beta_I - \beta_S). \tag{V.80}$$

Comparison of Eqs. (V.67, 71, and 80) yields finally:

$$R_{IS} = [2\eta_S \operatorname{Tr}(\hat{\mathscr{H}}_I'^2)]^{-1} \sum_{mu,nv} W_{mu,nv} (E_u' - E_v')^2, \tag{V.81}$$

and

$$R_{SI} = \frac{\eta_S \operatorname{Tr}(\hat{\mathscr{H}}_I'^2)}{\eta_I \operatorname{Tr}(\hat{\mathscr{H}}_S'^2)} R_{IS} = R_{IS}/\varepsilon. \tag{V.82}$$

For the calculation of the rate constants R_{SI} and R_{IS} we use a combination of first order perturbation theory and the method of moments, which takes account of the dipolar coupling amongst the I spins [13, 94]. In order to simplify the problem we consider the system as consisting of equal volumes each of which contains only a single S spin and an equal number of I spins. The behaviour of each of these subsystems is typical for the entire system. If both rf fields are exactly at resonance, the Hamiltonian of each of these subsystems is given as

$$\hat{\mathscr{H}}' = \hat{\mathscr{H}}_I' + \hat{\mathscr{H}}_S' + \hat{\mathscr{H}}_{dIS}^0, \tag{V.83}$$

where

$$\hat{\mathscr{H}}_I' = -\sum_p \gamma_I \hbar B_{1I} \hat{I}_{x'p} + \tfrac{1}{2} \sum_{p,q \neq p} \phi_{pq} (3\hat{I}_{zp}\hat{I}_{zq} - \hat{I}_p \cdot \hat{I}_q), \tag{V.83a}$$

$$\hat{\mathscr{H}}_S' = -\gamma_S \hbar B_{1S} \hat{S}_{x'}, \tag{V.83b}$$

$$\hat{\mathscr{H}}_{dIS}^0 = \sum_k \chi_k \hat{I}_{zk} \hat{S}_{zk}, \tag{V.83c}$$

$$\phi_{pq} = \tfrac{1}{2} \gamma_I^2 \hbar^2 (\mu_0/4\pi) r_{pq}^{-3} (1 - 3\cos^2 \vartheta_{pq}), \tag{V.83d}$$

$$\chi_k = \gamma_I \gamma_S \hbar^2 (\mu_0/4\pi) r_k^{-3} (1 - 3\cos^2 \vartheta_k). \tag{V.83e}$$

If $\hat{\mathscr{H}}_{dIS}^0$ would vanish, the two systems would be decoupled. Energy exchange is possible because of this interaction. This yields

$$
\begin{aligned}
R_{IS} &= [2\eta_S \operatorname{Tr}(\hat{\mathscr{H}}_I'^2)]^{-1} (2\pi/\hbar) \sum_{mu,nv} |\langle m, u| \hat{\mathscr{H}}_{dIS}^0 |n, v\rangle|^2 (E_u' - E_v')^2 \\
&\quad \cdot \delta(E_n' - E_m' + E_v' - E_u') \\
&= [2\eta_S \operatorname{Tr}(\hat{\mathscr{H}}_I'^2)]^{-1} (2\pi/\hbar) \sum_{\substack{m,u,n,v, \\ k,k'}} \chi_k \chi_{k'} \langle m| \hat{I}_{zk} |n\rangle \langle n| \hat{I}_{zk'} |m\rangle \\
&\quad \cdot (E_u' - E_v')^2 \\
&\quad \times \langle v| \hat{S}_z |u\rangle \langle u| \hat{S}_z |v\rangle \, \delta(E_n' - E_m' + E_v' - E_u').
\end{aligned} \tag{V.84}
$$

Equation (V.84) can be simplified, because the S spins have a negligible interaction so that the resonance condition in the rotating frame is $|E'_u - E'_v| = \gamma_S \hbar B_{1S} = \hbar \omega_{1S}$. Thus

$$R_{IS} = \frac{2\pi}{\hbar} \frac{\gamma_S^2 \hbar^2 B_{1S}^2}{2\eta_S \operatorname{Tr}(\mathscr{H}_I'^2)} \sum_{\substack{m,n,u,v, \\ k,k'}} \chi_k \chi_{k'} \langle m| \hat{I}_{zk} |n\rangle \langle n| \hat{I}_{zk'} |m\rangle$$
$$\times \langle v| \hat{S}_z |u\rangle \langle u| \hat{S}_z |v\rangle \, \delta(E'_n - E'_m - \hbar\omega_{1S}) . \tag{V.85}$$

One can evaluate the traces to obtain

$$R_{IS} = \frac{\pi\hbar\gamma_S^2 B_{1S}^2 S(S+1)}{3 \operatorname{Tr}(\mathscr{H}_I'^2)} \sum_{m,n,k,k'} \chi_k \chi_{k'} \langle m| \hat{I}_{zk} |n\rangle \langle n| \hat{I}_{zk'} |m\rangle$$
$$\times \delta(E'_n - E'_m - \hbar\omega_{1S}) . \tag{V.86}$$

Up to this point we only used first order perturbation theory. For the further discussion we need information about the energy values E'_m and E'_n. For that purpose we define a shape function

$$g(\omega_{1S}) = \sum_{m,n,k,k'} \chi_k \chi_{k'} \langle n| \hat{I}_{zk} |m\rangle \langle m| \hat{I}_{zk'} |n\rangle \, \delta(E'_n - E'_m - \hbar\omega_{1S}), \tag{V.87}$$

and evaluate it using the method of moments. In principle it is possible to compute all the moments of the shape function, Eq. (V.87), and to determine $g(\omega_{1S})$ in this way. However, since the calculation of higher moments is very complicated, we assume a special shape, for instance Gaussian or exponential.

The zeroth and second moment of these lines can be equalized with the respective data of Eq. (V.87). Comparison with the experimental results must decide which test function is the most reasonable.

Let us first calculate the moments according to Eq. (V.87), restricting ourselves to the case of complete adiabatic demagnetization of the I spin system in the rotating frame. Only for this case there are enough experimental results for T_{IS} for a comparison with the theoretical results. (The case of $B_{1I} \gg B'_{loc}$ was discussed extensively by Lurie and Slichter [13].) Because of the δ-function in Eq. (V.87) the right hand side contributes to the moments only if $\hbar \omega_{1S} = E'_n - E'_m$. However, for each pair of states there is one value of ω_{1S} which satisfies this condition. Therefore we get

$$\int_{-\infty}^{+\infty} g(\omega_{1S}) \, d\omega_{1S} = \hbar^{-1} \sum_{n,m,k,k'} \chi_k \chi_{k'} \langle n| \hat{I}_{zk} |m\rangle \langle m| \hat{I}_{zk'} |n\rangle . \tag{V.88}$$

Equation (V.88) can be evaluated using the well known condition

$$\sum_{n,m} \langle n| \hat{I}_{zk} |m\rangle \langle m| \hat{I}_{zk'} |n\rangle = \sum_n \langle n| \hat{I}_{zk}^2 |n\rangle = \tfrac{1}{3} I(I+1)(2I+1)^{N_I} . \tag{V.89}$$

This gives

$$\int_{-\infty}^{+\infty} g(\omega_{1S})\, d\omega_{1S} = \tfrac{1}{3}\hbar^{-1} I(I+1)(2I+1)^{N_I} \sum_k \chi_k^2. \tag{V.90}$$

The second moment can be calculated using van Vleck's method:

$$\int_{-\infty}^{+\infty} \omega_{1S}^2 g(\omega_{1S})\, d\omega_{1S} = \sum_{n,m,k,k'} \int_{-\infty}^{+\infty} \omega_{1S}^2 \chi_k \chi_{k'} \langle n|\, \hat{I}_{zk}\, |m\rangle \langle m|\, \hat{I}_{zk'}\, |n\rangle$$

$$\cdot \delta(E_n' - E_m' - \hbar\omega_{1S})\, d\omega_{1S} \tag{V.91}$$

$$= -\hbar^{-3} \sum_{n,m,k,k'} \chi_k \chi_{k'} \langle n|\, [\hat{\mathscr{H}}_I', \hat{I}_{zk}]\, |m\rangle$$

$$\cdot \langle m|\, [\hat{\mathscr{H}}_I', \hat{I}_{zk'}]\, |n\rangle.$$

In the case of complete adiabatic demagnetization $\hat{\mathscr{H}}_I'$ consists only of the secular part of the I spin dipolar interaction,

$$\hat{\mathscr{H}}_I' = \hat{\mathscr{H}}_{dII}^0 = \tfrac{1}{2} \sum_{p,q\,\neq\,p} \phi_{pq}(3\hat{I}_{zp}\hat{I}_{zq} - \hat{I}_p \cdot \hat{I}_q). \tag{V.92}$$

Substitution of Eq. (V.92) into Eq. (V.91) and application of the commutation relations for the components of the angular momentum finally yields

$$\int_{-\infty}^{+\infty} \omega_{1S}^2 g(\omega_{1S})\, d\omega_{1S} = \tfrac{1}{4}\hbar^{-3} \sum_{\substack{n,m,k,k' \\ p\,\neq\,k,\, p'\,\neq\,k'}} \chi_k \chi_{k'} \langle n|\, 2\phi_{pk}(-\hat{I}_{yk}\hat{I}_{xp} + \hat{I}_{xk}\hat{I}_{yp})\, |m\rangle$$

$$\times \langle m|\, 2\phi_{p'k'}(-\hat{I}_{yk'}\hat{I}_{xp'} + \hat{I}_{xk'}\hat{I}_{yp'})\, |n\rangle \tag{V.93}$$

$$= \tfrac{2}{9}\hbar^{-3} I^2(I+1)^2 (2I+1)^{N_I} \sum_{k,\, p\,\neq\,k} \chi_k^2 \phi_{pk}^2.$$

Equation (V.93) can be simplified by use of the second moments of the resonance line as obtained by van Vleck:

$$\langle \Delta^2 \omega_{II}\rangle = 3\hbar^{-2} I(I+1) \sum_{p\,\neq\,k} \phi_{pk}^2, \tag{V.94a}$$

$$\langle \Delta^2 \omega_{SI}\rangle = \tfrac{1}{3}\hbar^{-2} I(I+1) \sum_k \chi_k^2. \tag{V.94b}$$

$\langle \Delta^2 \omega_{II}\rangle$ is the contribution of the I spins to the second moment of the I spin resonance line, $\langle \Delta^2 \omega_{SI}\rangle$ the contribution of the I spins to the second moment of the S spin resonance line.

Thus the final result is

$$\int_{-\infty}^{+\infty} \omega_{1S}^2 g(\omega_{1S})\, d\omega_{1S} = \tfrac{2}{9}\hbar(2I+1)^{N_I} \langle \Delta^2 \omega_{II}\rangle \langle \Delta^2 \omega_{SI}\rangle. \tag{V.95}$$

In order to compare this result with the experiment we now choose two test functions:

$$g_1(\omega_{1S}) = C_1 e^{-\omega_1^2 s/\omega_1^2} \,, \tag{V.96a}$$

and

$$g_2(\omega_{1S}) = C_2 e^{-|\omega_1 s/\omega_2|} \,. \tag{V.96b}$$

The Gauss function, Eq. (V.96a) is the most common assumption, however the experimental results justify the exponential form given by Eq. (V.96b). The moments of both test functions can be calculated readily:

$$\int_{-\infty}^{+\infty} g_1(\omega_{1S}) \, d\,\omega_{1S} = (\pi)^{1/2} C_1 \omega_1 \,, \tag{V.97a}$$

$$\int_{-\infty}^{+\infty} g_2(\omega_{1S}) \, d\,\omega_{1S} = 2 C_2 \omega_2 \,, \tag{V.97b}$$

$$\int_{-\infty}^{+\infty} \omega_{1S}^2 g_1(\omega_{1S}) \, d\,\omega_{1S} = \tfrac{1}{2} \pi^{1/2} C_1 \omega_1^3 \,, \tag{V.98a}$$

$$\int_{-\infty}^{+\infty} \omega_{1S}^2 g_2(\omega_{1S}) \, d\,\omega_{1S} = 4 C_2 \omega_2^3 \,. \tag{V.98b}$$

By comparison of Eqs. (V.97) and (V.98) with Eq. (V.93) we get

$$C_1 = \frac{3}{2} \hbar \frac{(2I+1)^{N_I}}{(\pi \langle \Delta^2 \omega_{II} \rangle)^{1/2}} \langle \Delta^2 \omega_{SI} \rangle \,, \tag{V.99a}$$

$$C_2 = \frac{3}{2} \hbar \frac{(2I+1)^{N_I}}{(\langle \Delta^2 \omega_{II} \rangle)^{1/2}} \langle \Delta^2 \omega_{SI} \rangle \,, \tag{V.99b}$$

$$\omega_1^2 = \tfrac{4}{9} \langle \Delta^2 \omega_{II} \rangle \,, \tag{V.99c}$$

$$\omega_2 = \tfrac{1}{3} (\langle \Delta^2 \omega_{II} \rangle)^{1/2} \,. \tag{V.99d}$$

Thus the respective results for a Gaussian and an exponential test function are:

$$T_{IS}^{-1} = \tfrac{3}{2} \sqrt{\pi} \left[1 + \frac{S(S+1)\,\gamma_S^2 B_{1S}^2 N_S}{I(I+1)\,\gamma_I^2 B_{\text{loc}}'^2 N_I} \right] \frac{\langle \Delta^2 \omega_{SI} \rangle}{(\langle \Delta^2 \omega_{II} \rangle)^{1/2}} \tag{V.100a}$$
$$\cdot \exp(-\tfrac{9}{4} \gamma_S^2 B_{1S}^2 / \langle \Delta^2 \omega_{II} \rangle)$$

$$T_{IS}^{-1} = \tfrac{3}{2} \pi \left[1 + \frac{S(S+1)\,\gamma_S^2 B_{1S}^2 N_S}{I(I+1)\,\gamma_I^2 B_{\text{loc}}'^2 N_I} \right] \frac{\langle \Delta^2 \omega_{SI} \rangle}{(\langle \Delta^2 \omega_{II} \rangle)^{1/2}} \tag{V.100b}$$
$$\cdot \exp(-3 \gamma_S B_{1S} / \sqrt{\langle \Delta^2 \omega_{II} \rangle}) \,.$$

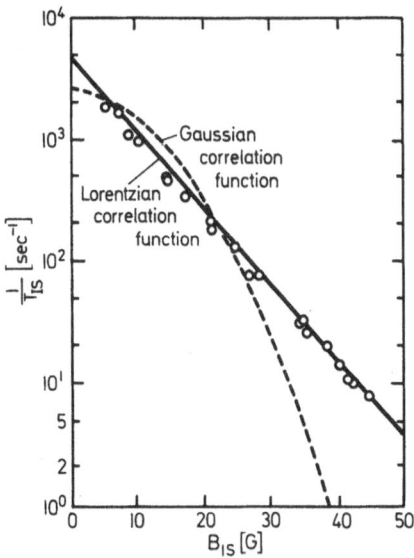

Fig. 17. Cross relaxation rate $1/T_{IS}$ in $CaF_2(I = {}^{19}F, S = {}^{43}Ca)$. The circles mark the experimental values [16]. The full and the dashed lines were calculated using Eqs. (V. 100 a) and (V. 100 b), respectively

This result was obtained by McArthur et al. [16] in a more formal way. They solved the equation of motion of the density matrix (master equation), and also included the case where ω_S is not exactly at resonance. T_{IS} is the time constant for the approach to a common temperature of the I and S spin system. This time constant can be measured by varying the length of the B_{1S} pulses. McArthur determined T_{IS} as a function of B_{1S} and tested Eq. (V.100) in this way. The sample was CaF_2, where the ^{43}Ca nuclei (0.13%) served as S spins and the ^{19}F nuclei (100%) as I spins. In Fig. 17 the experimental data are compared with the exchange times as calculated according to Eqs. (V.100a and b), respectively. B_0 was oriented parallel to [111] and for the calculation the following second moments have been used $\langle \Delta^2 \omega_{II} \rangle = 1.39 \cdot 10^9$ sec^{-2}, and $\langle \Delta^2 \omega_{SI} \rangle = 3.57 \cdot 10^7$ sec^{-2}. There is excellent agreement of the experimental data with the straight line which has been calculated using the exponential test function. That is, the correlation function in the rotating frame is Lorentzian and not Gaussian [12–14], as assumed in most cases.

Qualitatively similar results were found in LiF [17] with ^6Li (7.43%) as S spins and ^7Li (92.57%) as I spins (Fig. 18). Once again $1/T_{IS}$ decreases exponentially with B_{1S}. However, there is only qualitative agreement with the theoretical dependence according to Eq. (V.100b). Probably this discrepancy has the following reason: In CaF_2 the gyromagnetic

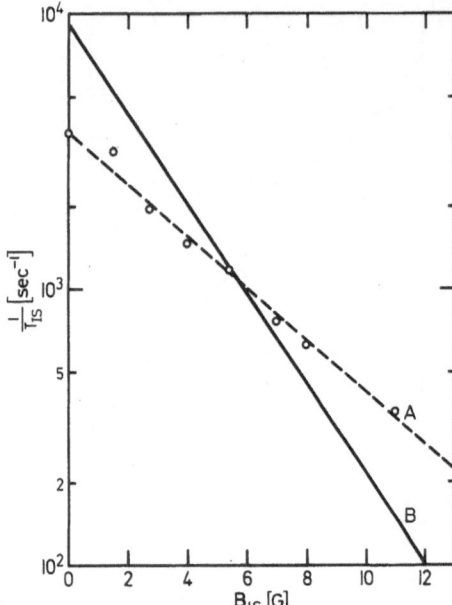

Fig. 18. Cross relaxation rate $1/T_{IS}$ in LiF ($I = {}^7$Li, $S = {}^6$Li). The circles mark the experimental values [17]. The full line was calculated using Eq. (V. 100b)

ratios of the S and I spins differ strongly ($\gamma_S : \gamma_I = 0.071$). Therefore $\hat{\mathscr{H}}^0_{dIS}$ is only a very small perturbation. On the other hand, the gyromagnetic ratios of ^{6}Li and ^{7}Li differ only by the small factor $1 : 2.64$. Therefore the restriction to first order perturbation theory is not justified any more. A similar fact has been observed earlier in the experiments by Bloembergen et al. and Pershan [94, 95]. The flipping of an S spin does not only flip a single I spin, but rather rearranges the whole dipolar system because of the interaction among the I spins, and also because of the interaction with the ^{19}F spins. However, it has not been possible to treat such processes theoretically in higher order perturbation theory.

VI. Spin Diffusion and Nuclear Double Resonance

VI.1. Observation of the Spin Diffusion Bottleneck

For a succesful DNMR experiment spin diffusion within the I spin system is necessary. In first approximation the S spins interact only with the nearest neighboring I spins, and spin diffusion has to conduct their high temperature to the entire I spin system. Spin diffusion is important for nuclear relaxation caused by paramagnetic centers (and for dynamic nuclear polarization). It has been discussed extensively in this context [76, 97]. Hartmann and Hahn already emphasized its importance. In Section IV.3.C it was indicated that second order quadrupole effects might inhibit the spin diffusion and thus might cause a bottleneck between the two spin systems. In I spin systems without quadrupole interaction the coupling among the I spins is usually stronger than the coupling between the S and I systems. In this case no diffusion bottleneck is expected. DNMR is useful for the investigation of spin diffusion since the heating rate of the S spin system can be varied and the dynamics of spin diffusion can be studied in this way.

In order to explain these experiments the influence of slow spin diffusion in a conventional NMR experiment shall be discussed. Suppose, the sample has been exposed to B_{1S} for a time T_h. Then the I spin temperature depends on the distant r from the spin S, if spin diffusion is slow. The I spin signal at the end of the DNMR experiment is proportional to the spatial average of the magnetization, i.e. to the spatial average of the inverse I spin temperature [36]:

$$M_I(t = T_h) = C_I \mu_0^{-1} B_{1I} \left(\frac{B_{loc}'^2}{B_{1I}^2 + B_{loc}'^2} \right)^{1/2} \int d\,V / \theta_I'(r, t = T_h). \tag{VI.1}$$

If, subsequently, one leaves the I spins in the demagnetized state for a delay time t_d, they will reach a uniform spin temperature. However, the average magnetization does not change during this process. This can be shown using the following simple argument: Apart from spin lattice relaxation effects, which can be corrected for in the experimental results, the internal energy of the I spin system during the delay time is constant. Thus we have

$$\langle E_I' \rangle = - C_I \mu_0^{-1} (B_{1I}^2 + B_{loc}'^2) \int d\,V / \theta_I'(r, t = T_h) = - C_I \mu_0^{-1} (B_{1I}^2 + B_{loc}'^2)$$

$$\cdot \int d\,V / \theta_I'(r, t = T_h + t_d) \tag{VI.2}$$

from which follows that

$$M_I(t = T_h) = M_I(t = T_h + t_d). \tag{VI.3}$$

Consequently, no information about the spatial distribution of the I spin temperature is obtained by waiting after switching off B_{1S} before inspecting the FID signal.

How can one observe the build-up of a nonuniform I spin temperature, caused by a diffusion bottleneck? Assuming that the S spins represent a continuous source of heat flowing into the I system, the I nuclei close to the impurity will rise to a spin temperature higher than average, if the spin diffusion is not fast enough to communicate this heat flow to

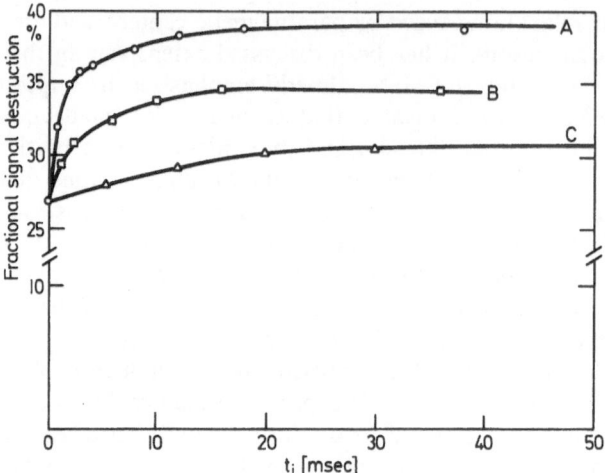

Fig. 19. Double resonance sensitivity with intermittent heating in NaCl:Ag (0.03 %). The fractional destruction of I spin magnetization is plotted versus the interruption time t_i between the heating bursts. The total heating time was 1.6 sec, the length of the individual burst was 2, 4, and 20 msec for curves A, B, and C, respectively

all the I nuclei. The S spins then can couple only to more distant I spins and consequently the double resonance sensitivity decreases. If the heat flow were interrupted, the I spins would reach uniform spin temperature after a certain time. If then the heat source were turned on again, the full initial double resonance sensitivity would be observed. Consequently, in order to observe the build-up of a nonuniform spin temperature one has to use groups of heating pulses separated by some interruption time t_i, instead of continuous heating.

In Fig. 19 the result of such an experiment is shown [18]. It was obtained using NaCl crystals doped with 0.03 % AgCl. The $\langle 200 \rangle$ neighbors of the impurity served as S spins [22], the unperturbed ^{23}Na nuclei as I spins. This system was chosen because it was expected that second order quadrupole effects inhibit the spin diffusion in the vicinity of the impurity, but still do not prevent the observation of the DNMR. In Fig. 19 the relative DNMR signal intensity is plotted as a function of t_i after irradiation of B_{1S} for a total time of 1.6 sec ($B_{1S} = 1.2$ G). During the double resonance process the I spins were totally demagnetized in the rotating frame ($B_{1I} = 0$). B_{1S} was irradiated intermitently with pulse lengths of 2 msec, 4 msec and 20 msec in curves A, B and C, respectively. The phase of B_{1S} was shifted by 180° at a rate of 1 kHz. During these pulses the S spin temperature was approximately infinite. For $t_i = 0$ the 3 curves converge to the same value. However, for short pulse lengths (curve A) and long interruption times, t_i, the DNMR signal increased by approximately 50% as compared to continuous irradiation of B_{1S}. For longer B_{1S} bursts a less effective increase of the double resonance sensitivity was observed, which indicated a gradual build-up of an inhomogeneous I spin temperature during a single heating burst.

For the interpretation of the results of Fig. 19 a time constant τ_{IS} can be defined which one would observe for the increase of the I spin temperature, if no diffusion bottleneck would exist. For the data at continuous heating one has to add a second time constant τ_{II} which takes account of the slow spin diffusion. Both time constants can be determined from the experiments with intermittent irradiation of B_{1S} giving

$$\tau_{IS} = 1.1 \, c_S^{-1} \text{ msec},$$

and (VI.4)

$$(\tau_{IS} + \tau_{II}) = 1.67 \, c_S^{-1} \text{ msec},$$

where c_S is the S spin concentration.

VI.2. Perturbation Theorie of the Diffusion Bottleneck

In order to estimate the influence of spin diffusion we begin with the diffusion equation as given by Bloembergen [76]

$$\partial \beta_I(r, t)/\partial t = D \nabla^2 \beta_I(r, t) - W_{IS}(r) \, (\beta_I(r, t) - \beta_S), \tag{VI.5}$$

where $\beta_I(r, t)$ is the inverse I spin temperature defined by

$$\beta_I(r, t) = 1/k\theta_I'(r, t). \tag{VI.6}$$

Since in a DNMR experiment the S spins are kept at high temperature, β_S is approximated as $\beta_S = 0$.

The first term at the right hand side of Eq. (VI.5) describes the spin diffusion within the I spin system. Since the diffusion approach is valid only on a scale large compared to the lattice constant, the diffusion constant D is assumed to be zero in a small sphere of radius b, where b is probably on the order of the nearest I spin neighbor distance. The second term at the right hand side of Eq. (VI.5) gives the local heat inflow caused by the direct coupling between an S spin and an I spin at a distance r from the S spin. $W_{IS}(r)$ is the probability for a mutual flipping process of such a pair of I and S spins. Although the S spins communicate with the I spins through their mutual dipolar interaction the resulting coupling rate is approximated by a simple radial function

$$W_{IS}(r) = K/r^6, \tag{VI.7}$$

neglecting the angular dependence of the interaction. Because of the r^{-6} dependence of the coupling rate, the S spins will efficiently interact only with close neighbors. Consequently in zeroth order approximation the diffusion problem can be solved using the boundary conditions $(\partial \beta_I / \partial r)_{r=b} = 0$ (since $D = 0$ for $r \leq b$), and $(\partial \beta_I / \partial r)_{r=B} = 0$, since there is no net heat flow across this boundary. If the direct coupling between the I and S spins can be neglected, the solutions of the diffusion equation are given as

$$\beta_{k_0}(r, t) = (3/4\pi B^3)^{1/2} = |k_0\rangle, \tag{VI.8a}$$

and

$$\beta_{k_n}(r, t) = (2\pi B)^{-\frac{1}{2}} r^{-1} \sin(k_n r) \exp - (\alpha_n^2 \pi^2 D t / B^2) = |k_n\rangle e^{-\mu_n t}. \tag{VI.8b}$$

In this equations the following abbreviation was used:

$$\mu_n = k_n^2 D = \alpha_n^2 \pi^2 D / B^2, \tag{VI.9}$$

where the α_n's are numerical factors which have the limits $n \leq \alpha_n \leq n + \frac{1}{2}$. If now $W_{IS}(r)$ is added as a small perturbation the solution of the total differential equation is slightly changed

$$\beta_{q_m}(r, t) = |q_m\rangle e^{-\lambda_m t}. \tag{VI.10}$$

Equation (VI.10) defines the radial eigenfunctions $|q_m\rangle$ of the problem and the rate constants λ_m. Experimentally one observes the spatial average of $\beta_I(r, t)$. A simple calculation yields

$$\overline{\beta}_I(t) = \beta_i \sum_m |\langle k_0 | q_m \rangle|^2 e^{-\lambda_m t}, \tag{VI.11}$$

where the λ_m's can be estimated using second order perturbation theory:

$$\lambda_m = \mu_m + \langle k_m | W_{IS}(r) | k_m \rangle + \sum_n \frac{\langle k_m | W_{IS}(r) | k_n \rangle \langle k_n | W_{IS}(r) | k_m \rangle}{\mu_m - \mu_n}. \tag{VI.12}$$

β_i is the initial inverse I spin temperature before irradiation of B_{1S}. At this time β_I is constant:

$$\overline{\beta}_I(t = 0) = \beta_i |k_0\rangle. \tag{VI.13}$$

After switching on B_{1S}, $\overline{\beta}_I$ becomes time dependent according to Eq. (VI.11). For small concentrations of S it is easy to show that the terms with $m > 0$ are negligible. One then obtains for the average of the I spin magnetization:

$$\overline{\beta}_I(t) = \beta_i e^{-\lambda_0 t} = \beta_i e^{-t/\tau_0}, \tag{VI.14}$$

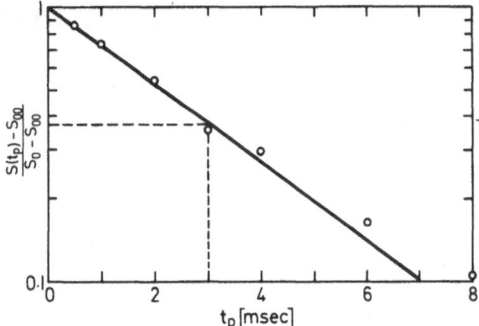

Fig. 20. Determination of the cross relaxation rate $1/T_{IS}$ in NaCl:Ag. S_0 is the I spin resonance signal observed without B_{1S}, S_∞ the signal obtained at very long B_{1S} pulses (15 msec, $B_{1S} = 1.2$ G) and $S(t_p)$ the signal at a pulse length t_p

where τ_0 can be estimated to be

$$\tau_0 = \tfrac{3}{8} d^6 (\sqrt{2}\, n^* \pi c_S \cdot K)^{-1} \left[1 + (7Dd^4/8K - 1)^{-1}\right]. \tag{VI.15}$$

n^* is the number of I spins in the unit cell. In the case under consideration (NaCl:Ag) $n^* = 4$. The first term at the right hand side of Eq. (VI.15) gives the time constant which one would observe, if spin diffusion were very rapid, that is τ_{IS}. The second term increases the time constant, if D is small. τ_0 is therefore identical with $(\tau_{IS} + \tau_{II})$. In order to determine τ_{IS} and τ_{II} the constants D and K are required. They can be estimated using the theories of Bloembergen [76] and Khutsishvili [97]. If these theories are extended to spins $I = \tfrac{3}{2}$ one obtains

$$D \approx 200\, d^2\ \sec^{-1}, \tag{VI.16}$$

where d is the full lattice constant of NaCl. It is more difficult to estimate K. One should apply the methods discussed in Section V.4. However, since the coupling, between the I and S spins is nearly equal to the $I - I$ coupling perturbation theory is not justified. Furthermore, in Section V.5 quadrupole interactions are disregarded. Therefore, it is more appropriate to determine K semiempirically using the measured value of the time constant T_{IS} for the temperature exchange between the I and S spin system after a single heating of the S spins. T_{IS} has been measured in the same way as the values of T_{IS} in CaF$_2$ and LiF in Fig. 17 and Fig. 18, respectively. In Fig. 20 the fraction $(S(t_p) - S_\infty)/(S_0 - S_\infty)$ is plotted as a function of the pulse length $t_p(B_{1S} = 1.2$ G). S_0 is the I spin resonance signal observed without B_{1S}, S_∞ the signal which was obtained at very long B_{1S} pulses (15 msec), and $S(t_p)$ the signal at a pulse length t_p. Fig. 20 yields $T_{IS} = 3$ msec. According to Section V.4

$$1/T_{IS} = R_{IS}(1 + 1/\varepsilon), \tag{VI.17}$$

where $1/\varepsilon$ is the ratio of the specific heats of the I and the S spin system. On the other hand Eqs. (VI.5) and (VI.7) yield

$$R_{IS} = (64\pi c_S/3\sqrt{2})\,(K/d^6)\,. \tag{VI.18}$$

If one substitutes the value of ε,

$$\varepsilon = \tfrac{3}{5}c_S\,\frac{\gamma_S^2 B_{1S}^2}{\gamma_I^2(B_{loc}'^2 + B_{1I}^2)}\,, \tag{VI.19}$$

one obtains finally

$$K/d^6 = \tfrac{9}{320}(\sqrt{2}/\pi)\,\frac{\gamma_S^2 B_{1S}^2}{\gamma_I^2(B_{1I}^2 + B_{loc}'^2)}\,T_{IS}^{-1}\,. \tag{VI.20}$$

Equation (VI.20) gives in the case under consideration ($B_{1S} = 1.2$ G, $B_{loc}' = 0.578$ G, $B_{1I} = 0$, $\gamma_S = \gamma_I$):

$$K/d^6 = 18.3\ \text{sec}^{-1}\,. \tag{VI.21}$$

Using this value one gets finally from Eq. (VI.15) the theoretical values

$$\tau_{IS} = 1.15\ \text{msec}\,,\quad\text{and}\quad (\tau_{IS} + \tau_{II}) = 1.3\ \text{msec}\,. \tag{VI.22}$$

Regarding the rough approximations, which have been employed in this calculation, the agreement with the experimental result is quite satisfying. The fact that the theoretical τ_{IS} is very close to its experimental value suggests that the estimate of D is too high at least for the region near the impurity, perhaps because the model has simplified many of the details of the real situation. Moran and Lang [19] examined the diffusion bottleneck in a similar way by studying the dynamics of DNMR in LiF (^{6}Li $= S$, ^{7}Li $= I$). The approximations are essentially the same as those used in the perturbation theory in this section.

The experiments in NaCl:Ag proof the existence of a diffusion bottleneck. It does not seriously limit the double resonance sensitivity in this case. However, the success of a DNMR experiment can be prevented by diffusion bottlenecks, if second order quadrupole effects cannot be neglected.

VII. Conclusion: Information Obtained by DNMR

DNMR can be employed for the investigation of solids. The strong dipolar interaction between the I and the S spins makes a DNMR experiment feasible. In liquids the rapid motion cancels this interaction. Consequently DNMR does not yield information as obtained with NMR of liquids since the resonance lines in solids are too wide (for instance, information about chemical shifts obtained by high resolution NMR; see, however, appendix 3).

The most important application of DNMR is the investigation of diamagnetic defects. It yields detailed information about the electrical field gradients, for instance, in the vicinity of impurities in alkali fluorides. This, in turn, provides knowledge about the electronic structure of ionic crystals (e.g. about the overlap of the wave functions of neighboring ions and about the possible existence of covalent binding). If the abundant spins possess a nuclear quadrupole moment, the high field DNMR may fail to work in the study of electrical field gradients near impurities. The most probable reason for this failure is a diffusion bottleneck caused by second order quadrupole effects. Zero field quadrupole double resonance makes it possible to investigate such systems, but the interpretation of the spectra obtained by this method may be rather difficult. It seems feasible to study the structure of molecules with double resonance methods; the rare isotopes ^{2}H, ^{17}O and ^{31}S, for instance, render themselves as probes for electric field gradients within the molecules. Presumably their resonance frequencies may be measured with double resonance methods. Using NQDR even powder samples can be studied. However, in zero magnetic field the dipolar coupling of nuclei with integer spins may be quenched by nonaxial symmetric electric field gradients [98]. Therefore NQDR may fail to work, for instance with ^{2}H, since it may be decoupled from the I spins (^{1}H).

Finally DNMR furnishs the possibility to explore the dynamics of interacting spin systems under the influence of strong rf fields. This provides valuable information for the theoretical understanding of the nature of the interaction. On the other hand, there are practical consequences, since the results can be used for the understanding of dynamical nuclear polarization.

Acknowledgements. The author is grateful to Prof. H. C. Wolf for the continuous support and interest in the progress of this work. Special thanks are due to Prof. C. P. Slichter for many stimulating discussions and valuable suggestions. The advice of Prof. H. Seidel is gratefully acknowledged.

Appendices

Appendix 1: Double Resonance Condition for S Spins with Quadrupole Interaction

The double resonance condition, Eq. (II.8) was derived for nuclei which are exposed to magnetic interactions exclusively. With quadrupole interaction it must be slightly modified. Consider the energy levels of the S spins in the rotating frame. In first order approximation the Hamil-

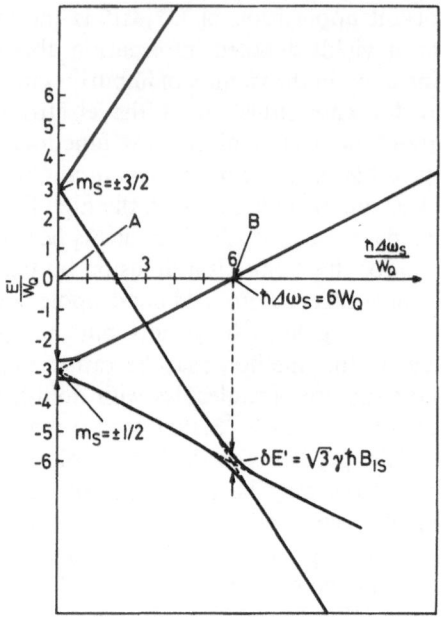

Fig. 21. Rotating frame energy levels of a spin $S = \frac{3}{2}$ in a high magnetic field with an axially symmetric quadrupole interaction $\hat{\mathscr{H}}_Q = W_Q[3\hat{S}_z^2 - S(S+1)]$

tonian is given as

$$\hat{\mathscr{H}}_S' = W_Q(3\hat{S}_z^2 - \hat{S}^2) + \hbar\Delta\omega_S\hat{S}_z - \gamma_S\hbar B_{1S}\hat{S}_{x'}, \tag{A.1}$$

where W_Q is defined by Eq. (IV.20) and $\Delta\omega_S = \omega_S - \gamma_S B_0$. If $\hbar\Delta\omega_S \gg \gamma_S\hbar B_{1S}$, the latter term can be neglected in zeroth order approximation. The energy levels in the rotating frame are then

$$E_1^{(0)} = W_Q[3m_S^2 - S(S+1)] + \hbar\Delta\omega_S m_S. \tag{A.2}$$

This is illustrated in Fig. 21 for a spin $S = \frac{3}{2}$. If, however, ω_S satisfies the resonance condition $\omega_S = \gamma_S B_0 - 6W_Q m_q/\hbar$ (see Section IV.4.A), the energy levels in the rotating frame are degenerate (A and B in Fig. 21). In this case the rf term must not be neglected.

It lifts the degeneracy and causes a splitting of the levels in the rotating frame

$$\Delta E_{1S} = \sqrt{S(S+1) - \frac{1}{4}(4m_q^2 - 1)}\, \gamma_S\hbar B_{1S}. \tag{A.3}$$

The double resonance condition Eq. (II.8) therefore has to be replaced by

$$\gamma_I B_{1I} = \sqrt{S(S+1) - \frac{1}{4}(4m_q^2 - 1)}\, \gamma_S B_{1S}. \tag{A.4}$$

Appendix 2: Nuclear Double Resonance Using Three Different Spin Species

An interesting modification of the DNMR method was introduced by Schwab and Hahn [99]. In the experiments, discussed so far, always two nuclear species I and S were involved in the double resonance process. For a successful DNMR experiment the lifetime of the ordered state, T_1^*, has to be as long as possible. This, however, requires a long spin lattice relaxation time in the laboratory frame. Following the FID for the measurement, the I spin magnetization is lost irreversibly and recovers with the long time constant, T_1. Spin locking is possible only after the equilibrium magnetization in the laboratory frame is regained. Therefore double resonance spectra can be scanned only slowly. If there is a third nuclear spin species, X, with a short spin lattice relaxation time, T_{1X}, this slow cooling process can be avoided by coupling the X and the I spin system. Schwab and Hahn used this technique for a sensitive detection of rare nuclei in paradichlorobenzene. The zero field quadrupole resonance frequency of the ^{35}Cl nuclei (X spins) is 34.78 MHz, the spin lattice relaxation time, $T_{1X} = 0.39$ sec at $77°$ K. The spin lattice relaxation time of protons (I spins) of the same resonance frequency amounts to several hours. However, both systems can be coupled, if the X spins are phase locked and if the double resonance condition $\gamma_X B_{1X} \approx \gamma_I B_{loc}$ is satisfied [88]. Following the establishment of thermal equilibrium of the two spin systems, B_{1X} can be switched off. This exchange process can be repeated several times within a few seconds. In this way the protons can be polarized to a degree equivalent to thermal equilibrium at 8100 G. The ordered I spin system can be used for a conventional DNMR experiment. Using this method Schwab and Hahn observed the NMR of the 1.1% naturally abundant ^{13}C nuclei at $B_0 = 130$ G. The detection of deuterium was possible only in enriched samples (12%) because of the strong overlap of the nonresonant absorption [44] of the protons with the deuterium resonance line. The remaining order of the I spin system can also be measured using the X spins. For, if the I spins are coupled to the X spins, the equilibrium temperature of the X spins is proportional to the I spin temperature. Therefore the signal amplitude during the FID of the spin locked ^{35}Cl nuclei measures the I spin temperature.

Appendix 3: High Resolution DNMR in Solids

The information obtained from DNMR experiments, as discussed so far, is severely limited by the large width of the resonance line due to the dipolar interaction. Recently new techniques for high resolution DNMR were reported [100–103]. These methods combine the established ideas of double resonance and spin decoupling. Spin decoupling using strong rf fields was discussed in great detail in a series of papers by Waugh and coworkers [104–109]. The application of this technique brings about the effective elimination of the normally predominant dipolar interactions between nuclear spins in solids. Because of the resulting reduction of the linewidth it is then possible to study weaker effects such as chemical shifts.

A typical high resolution DNMR experiment consists of the following four steps [103]:

1. Polarize the I spins in a high field.

2. Cool the I spins to a low spin temperature in the rotating frame.

3. Establish cross relaxation between the I and the S system.

4. Record the S spin FID signal, while decoupling the I spins by an appropriate B_{1I} pulse sequence [103, 108].

Using Fourier transform spectroscopy [110] it was possible to investigate the NMR of ^{13}C in natural abundance in 50 mg of solid adamantane, $C_{10}H_{16}$ [103]. Two lines of a width of about 40 Hz were observed, shifted (87.5 ± 1) ppm and (96.2 ± 1) ppm upfield from a reference of liquid $^{13}C_6H_6$ (proton decoupled). The residual linewidth is limited by nonideal experimental conditions. In spectra from ^{13}C in natural abundance a residual linewidth of about 5 Hz is expected, due to the dipolar coupling among the S spins.

It can be shown that this technique permits the detection of the NMR of ^{13}C in natural abundance at a signal to noise ratio of about 10^6. High resolution NMR of very rare spins is therefore feasible and may become a powerful tool for the study of chemical shifts of rare spins and structural investigations in solids.

References

1. Overhauser, A. W.: Phys. Rev. **92**, 411 (1953).
2. Hausser, K. H., Stehlik, D.: Advances in Magnetic Resonance, Vol. 3, p. 79. New York, London: Academic Press 1968.
3. Jeffries, C. D.: Dynamic Nuclear Interaction, New York: Wiley, Interscience 1963.
4. Abragam, A.: The Principles of Nuclear Magnetism, Chapter 9. London: Oxford University Press, 1961.
5. Kastler, A.: J. Phys. Radium **11**, 255 (1950).
6. Bernheim, R.: Optical Pumping. New York: W. Benjamin 1965.
7. Feher, G.: Phys. Rev. **103**, 834 (1956)
8. Seidel, H.: Habilitationsschrift, Stuttgart (1966).
9. Brossel, J., Kastler, A.: Compt. Rend. **229**, 1213 (1949).
10. Brossel, J., Bitter, F.: Phys. Rev. **86**, 308 (1952).
11. zu Putlitz, G.: Ergeb. Exakt. Naturw. **37**, 105 (1965).

12. Hartmann, S. R., Hahn, E. L.: Phys. Rev. **128**, 2042 (1962).
13. Lurie, F. M., Slichter, C. P.: Phys. Rev. **133**, A 1108 (1964).
14. Redfield, A. G.: Phys. Rev. **130**, 589 (1962).
15. Refield, A. G.: Phys. Rev. **98**, 1787 (1955).
16. Mc Arthur, D. A., Hahn, E. L., Walstedt, R. E.: Phys. Rev. **188**, 609 (1969).
17. Lang, D. V., Moran, P. R.: Phys. Rev. B **1**, 53 (1970).
18. Spencer, P. R., Schmid, H. D., Slichter, C. P.: Phys. Rev. B **1**, 2989 (1970)
19. Moran, P. R., Lang, D. V.: Phys. Rev. B **2**, 2360 (1970).
20. Fernelius, N. C.: Proc. XIV. Colloque Ampère, Ljubljana 1966, p. 497. Editor R. Blinc. Amsterdam: North Holland Publishing Co. 1967.
21. Slusher, R. E., Hahn, E. L.: Phys. Rev. **166**, 332 (1968).
22. Mallick, G. T., Schumacher, R. T.: Phys. Rev. **166**, 350 (1968).
23. Hartland, A.: Proc. Phys. Soc. A **304**, 361 (1968).
24. Nelson, K. F., Ohlsen, W. D.: Phys. Rev. **180**, 366 (1969)
25. Dick, B. G., Nelson, K. F.: Phys. Rev. **186**, 953 (1969).
26. Fernelius, N. C., Slichter, C. P.: Proc. XV. Colloque Ampère, Grenoble 1968. Editor P. Averbuch, p.347. Amsterdam: North Holland Publishing Co. 1969.
27. Minier, M.: Phys. Rev. **182**, 437 (1969).
28. Berthier, C., Minier, M., Segransan, P.: Proc. XVI. Colloque Ampère, Bucharest 1970, p. 853, Bucharest: Editor I. Ursu, 1971.
29. Tsutsumi, Y., Kunitomo, M., Terao, T., Hashi, T.: J. Phys. Soc. Japan **26**, 16 (1969).
30. Stehlik, D., Nordal, P. E.: Proc. XVI. Colloque Ampère, Bucharest 1970, p. 356, Bucharest: Editor I. Ursu, (1971).
31. Slichter, C. P.: Principles of Magnetic Resonance, Chapter 2, New York: Harper & Row Publ. 1963.
32. Bloch, F.: Phys. Rev. **70**, 460 (1946).
33. Lowe, I. J., Norberg, R. E.: Phys. Rev. **107**, 46 (1957).
34. Franz, J. R., Slichter, C. P.: Phys. Rev. **148**, 287 (1966).
35. Kittel, C.: Elementary Statistical Physics, Chapter 6. New York: John Wiley & Sons Inc. (1964).
36. Slichter, C. P., Holton, W. C.: Phys. Rev. **122**, 1701 (1961).
37. Jeener, J., Broekaert, P.: Phys. Rev. **157**, 232 (1967).
38. Jones, E. P., Hartmann, S. R.: Phys. Rev. Letters **22**, 867 (1969); Phys. Rev. B **6**, 757 (1972).
39. Cohen, M. H., Reif, F.: Solid State Phys. **5**, 321 (1957).
40. Das, T. P., Hahn, E. L.: Solid State Phys. Suppl. 1, 1 (1958).
41. Rowland, T. J.: Acta Met. 3, 74 (1955).
42. Bloembergen, N.: Report of the Conference on Defects in Crystalline Solids, Physical Society, London (1954).
43. Feld, B. T., Lamb, W. E.: Phys. Rev. **67**, 15 (1945).
44. Anderson, A. G., Hartmann, S. R.: Phys. Rev. **128**, 2023 (1962).
45. Cohen, M. H.: Phys. Rev. **96**, 1278 (1954).
46. Drain, L. E.: J. Phys. C (Proc. Phys. Soc.) **1**, 1690 (1968).
47. Minier, M.: Private Communication.
48. Blandin, A., Friedel, J.: J. Phys. Radium **21**, 689 (1960).
49. Kohn, W., Vosko, S. H.: Phys. Rev. **119**, 912 (1960).
50. Lucken, E. A. C.: Nuclear Quadrupole Coupling Constants. New York: Academic Press 1969.
51. Fukai, Y., Watanabe, K.: Phys. Rev. B **2**, 2353 (1970).
52. Fukai, Y.: Phys. Rev. **186**, 697 (1969).
53. Watson, R. E., Gossard, A. C., Yafet, Y.: Phys. Rev. **140**, A 375 (1965).
54. Leppelmeier, G. W., Hahn, E. L.: Phys. Rev. **142**, 179 (1966).

55. Abragam, A.: Ref. 4, Chapter 7.
56. Andersson, L.O.: Arkiv Fysik **40**, 71 (1969).
57. Andersson, L.O.: Proc. Brit. Cerem. Soc. **9**, 83 (1967).
58. Andersson, L.O.: Helv. Phys. Acta **41**, 414 (1968).
59. Kawamura, H., Otsuka, E., Ishiwatari, K.: J. Phys. Soc. Japan **11**, 1064 (1956).
60. Andersson, L.O., Forslind, E.: Arkiv Fysik **28**, 49 (1965).
61. Ohlsen, W.D., Melich, M.E.: Phys. Rev. **144**, 240 (1966).
62. Fukai, Y.: J. Phys. Soc. Japan **18**, 1413 (1963); **18**, 1580 (1963); **19**, 175 (1964).
63. Kornfeld, M.I., Lemanov, V.V.: Sov. Phys.-JETP **16**, 1427 (1963).
64. Eshelby, J.D.: Solid State Phys. **3**, 79 (1956).
65. Shulman, R.G., Wyluda, B.J., Anderson, P.W.: Phys. Rev. **107**, 953 (1957).
66. Dick, B.G., Das, T.P.: Phys. Rev. **127**, 1053 (1962).
67. Das, T.P., Dick, B.G.: Phys. Rev. **127**, 1063 (1962).
68. Dick, B.G.: Phys. Rev. **145**, 609 (1966).
69. Tessman, J.R., Kahn, A.H., Shockley, W.: Phys. Rev. **92**, 890 (1953).
70. Das, T.P., Karplus, M.: J. Chem. Phys. **42**, 2885 (1965).
71. Spaeth, J.M.: Z. Phys. **192**, 107 (1966).
72. Kersten, R.:Dissertation, Stuttgart (1970).
73. Ikenberry, D., Das, T.P.: Phys. Rev. **184**, 989 (1969).
74. Satoh, M., Spencer, P.R., Slichter, C.P.: J. Phys. Soc. Japan **22**, 666 (1967).
75. Spencer, P.R.: Private Communication
76. Bloembergen, N.: Physica **15**, 386 (1949).
77. Lee, M., Goldburg, W.I.: Phys. Rev. **140**, A 1261 (1965).
78. Blinc, R.: Advances in Magnetic Resonance, Vol. 3, p. 141. Editor J. S. Waugh, New York-London: Academic Press 1968.
79. Bonera, G., Borsa, F., Rigamonti, A.: Proc. XVI. Colloque Ampère, Bucharest 1970, p. 349, Bucharest: Editor I. Ursu, (1971). – A. Rigamonti, Phys. Rev. Letters **19**, 436 (1967).
80. Blinc, R., Zumer, S., Lahajnar, G.: Phys. Rev. B **1**, 4456 (1970).
81. Abragam, A.: Ref. 4, Chapters 5 and 12.
82. Goldman, M.: Spin Temperature and Nuclear Magnetic Resonance in Solids. London: Oxford University Press 1970.
83. Pound, R.V.: Phys. Rev. **81**, 156 (1951).
84. Slichter, C.P.: Ref. 31, p. 53.
85. Abragam, A.: Ref. 4, Chapter 4.
86. Goldburg, W.I.: Phys. Rev. **128**, 1554 (1962).
87. Bloembergen, N., Sorokin, P.P.: Phys. Rev. **110**, 865 (1958).
88. Goldman, M., Landesman, A.: Phys. Rev. **132**, 610 (1963).
89. Van Vleck, J.H.: Phys. Rev. **74**, 1168 (1948).
90. Slichter, C.P.: Ref. 31, p. 26.
91. Slichter, C.P.: Ref. 31, Chapter 2.6.
92. Abragam, A., Proctor, W.G.: Phys. Rev. **109**, 1441 (1958).
93. Schumacher, R.T.: Phys. Rev. **112**, 837 (1958).
94. Bloembergen N., Shapiro, S., Pershan, P.S., Artmann, J.O.: Phys. Rev. **114**, 445 (1959).
95. Pershan, P.S.: Phys. Rev. **117**, 109 (1960).
96. Goldburg, W.I.: Phys. Rev. **115**, 48 (1959).
97. Khutsishvili, G.R.: Sov. Phys.-Uspekhi **8**, 743 (1966).
98. Leppelmeier, G.W., Hahn, E.L.: Phys. Rev. **141**, 724 (1966).
99. Schwab, M., Hahn, E.L.: J. Chem. Phys. **52**, 3152 (1970).
100. Mansfield, P., Grannell, P.K.: J. Phys. C.: Solid State Phys. **4**, L 197 (1971).
101. Bleich, H.E., Redfield, A.G.: J. Chem. Phys. **55**, 5405 (1971).

102. Yannoni, C. S., Bleich, H. E.: J. Chem. Phys. **55**, 5406 (1971).
103. Pines, A., Gibby, M. G., Waugh, J. S.: J. Chem. Phys. **56**, 1776 (1972).
104. Sarles, L. R., Cotts, R. M.: Phys. Rev. **111**, 853 (1958).
105. Waugh, J. S., Wang, C. H., Huber, L. M., Vold, R. L.: J. Chem. Phys. **48**, 662 (1968).
106. Waugh, J. S., Huber, L. M., Haeberlen, U.: Phys. Rev. Letters, **20**, 180 (1968).
107. Haeberlen, U., Waugh, J. S.: Phys. Rev. **175**, 453 (1968).
108. Mehring, M., Pines, A., Rhim, W. K., Waugh, J. S.: J. Chem. Phys. **54**, 3239 (1971).
109. Gibby, M. G., Pines, A., Rhim, W. K., Waugh, J. S.: J. Chem. Phys. **56**, 991 (1972).
110. Ernst, R. R., Anderson, W. A.: Rev. Sci. Instr. **37**, 93 (1966).

Dankward Schmid
Physikalisches Institut der Universität
D-7000 Stuttgart 1
Federal Republic of Germany

Vibrational Spectra of Electron and
Hydrogen Centers in Ionic Crystals

D. BÄUERLE

Contents

1. Introduction

The fundamental role played by lattice vibrations in nearly all fields of solid state physics, be it in connection with absorption of elastic or electromagnetic waves, superconductivity, or ferroelectricity, has stimulated a growing interest in lattice vibrational spectra of pure and perturbed crystal lattices. Such vibrations can be studied directly be means of inelastic neutron scattering which is to be prefered for the study of bulk vibrational modes, and by infrared and Raman spectroscopy which, because of their inherent sensitivity to relatively low defect concentrations, are excellent techniques for investigating the defect induced response of a crystal lattice. Such defects, which may be foreign atoms, ions, or molecules, but which may also be dislocations, voids, and so on, are interesting because they lower the point symmetry of the lattice, thereby relaxing the selection rules for photon-phonon interaction processes, and thus allowing the optical activation of (somewhat perturbed) host lattice vibrations which are inactive in the unperturbed crystal.

It is the aim of this work to give a survey on the infrared and Raman spectra of some color centers in ionic crystals, particularly for those whose mass is very small when compared with the mass of the host lattice ions. Electron, hydrogen, and vacancy centers, defects which have this property, are of special interest for the clarification of lattice dynamical problems. The reason is that the atomic and electronic structure of these centers is well established from their electronic absorption spectra, and from the analysis of electron-paramagnetic resonance (EPR), and/or electron nuclear double resonance (ENDOR) data. Furthermore, some lattice dynamical properties of these centers were already known. They were derived indirectly from the shape of the electronic absorption lines which show strong temperature – dependent broadening due to electron – phonon coupling. Extensive investigations of this kind were performed for the electronic absorption of F centers in alkali halides. In addition, electron and hydrogen centers can be transformed by simple and directly controlable photochemical reactions, thus allowing one to study, in a well defined way, point defects of various potentials, charges, and point symmetry. The experimental and theoretical analysis of the infrared and Raman spectra of these defects yields detailed information on the site symmetry and local bonding of the defect, and on the host latticeband modes.

The main features observed in the infrared and Raman spectra of the special defects discussed in this work are representative of the vibrational properties of defective crystal lattices in general.

2. General Remarks

2.1. Formation, Structure, and Electronic Absorption of Electron and Hydrogen Centers

The electron and hydrogen centers in alkali halides whose vibrational properties will be investigated in this work are shown schematically in Fig. 1. The following remarks on the production, atomic structure, and the electronic absorption of these centers are of an introductory nature. For more details, refer to special literature on this topic [1–4].

The most well known electron center is the F center, which is a trapped electron on an anion lattice site. In alkali halides, F centers can be produced by heating the crystal in alkali vapor (additive coloration), by electron injection (e.g. electrolytic coloration), by X-irradiation or by a photochemical reaction from U centers (see Fig. 1). F centers give rise to an electronic absorption in the visible spectral region (in KBr near 2.06 eV at $T = 4.2$ K). Irradiating with light into the visible F band at low temperatures results in the formation of F' centers (two electrons in an anion vacancy) and anion vacancies (α centers). The electronic absorption of F' centers is located in the visible to near infrared region (in KBr near 1.77 eV at 170 K) while α centers absorb in the near ultra-violet (in KBr near 6.2 eV at 90 K). Irradiating the F band with visible light at room temperature leads to an association of F centers into pairs (M centers), trimers (R centers), or even more complex aggregates. The

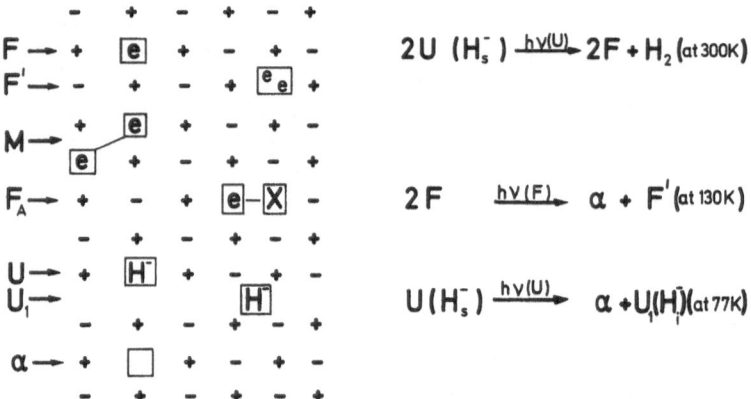

Fig. 1. Left side: Structural models for some electron and hydrogen centers in alkali halides. F center: one electron in an anion vacancy, F' center: two electrons in an anion vacancy, M center: two neighboring electrons in [110] or equivalent directions, F_A center: one electron associated with foreign alkali ion of smaller size, $U(H_s^-, D_s^-)$ center: substitutional hydrogen ion, $U_1(H_i^-, D_i^-)$ center: interstitial hydrogen ion, α center: anion vacancy. Right Side: Photochemical reactions. Energy $h\nu$ refers to electronic bands

electronic absorption of these centers is located in the visible to near infrared region. An F center associated with one foreign alkali ion of smaller size in the first shell is called an F_A center [in analogy to Fig. 32, case (b)].

Since the point symmetry in the F_A configuration is reduced to C_{4v}, the threefold degenerate first excited state of the F center splits twofold. F_A centers therefore give rise to two electronic transitions which are located near the F center transition (in KBr for F_A(Na) 2.07 and 1.90 eV, respectively at 4 K).

Negative hydrogen ions placed substitutionally on anion sites are called U centers, because they give rise to an electronic transition in the near ultraviolet region (in KBr near 5.51 eV at $T = 6$ K). U centers can be produced by treating additively coloured crystals under hydrogen atmosphere at about 200 K below the melting point, or by doping the melt with alkali hydride. U centers can be converted by UV or X-irradiation at room temperature into F centers and interstitial hydrogen molecules (see Fig. 1). Irradiating at liquid nitrogen temperature yields interstitial hydrogen centers, U_1 centers, and anion vacancies (generally this process leads simultaneously to the production of a small amount of F centers according to the first reaction shown in Fig. 1). The electronic U_1 band is very broad even at low temperatures (maximum at about 4.6 eV in KBr at 55 K).

In alkaline earth fluorides, defects which possess the structure of the F center can be produced by X-ray irradiation or by additive coloration when heating samples in the presence of aluminium metal (at about 800 K). Treating additive colored samples under hydrogen atmosphere (at about 1100 K) results in the production of substitutional hydrogen ions [5]. The formation of substitutional hydrogen atoms by photochemical reactions has been established as well [6]. The production of interstitial hydrogen ions and hydrogen atoms was established in rare earth ion doped CaF_2 (see Sections 3.1 and 3.3) [7].

2.2. Selection Rules for Infrared and Raman Spectra

In pure, ideal crystals the interaction of elementary excitations with light demands conservation of (quasi-) momentum, This follows from the translational symmetry of crystal lattices. Because the wavelength of light is very large when compared with the crystal unit cell, only lattice vibrations with wave vector $k \approx 0$ can be directly excited. In ionic crystals vibrations near the center of the Brillouin zone are dipole active if they belong to transverse optical branches in the phonon dispersion curves. The corresponding oscillators are sometimes called "dispersion oscillators" or "reststrahl oscillators". Dispersion oscillators give rise to optical absorption in the middle to far infrared frequency region.

When point defects are introduced into the crystal, the lattice vibrations can no longer be classified according to the wavevector, but only according to the point symmetry of the defect. For the interaction with light the k conservation law is no longer valid, i.e. all phonons may become infrared active in the *one* phonon absorption. The corresponding dipole moment is determined by the changed eigenvectors and the additional dynamical charge of the defect.

From atomic physics it is well known that optical transitions will be allowed only for certain combinations of quantum states. The selection rules can be expressed most easily in group theoretical terms (for details see Refs. [8–11]). Assume that the lattice dynamical problem is described by a Hamiltonian H which is invariant under operations $\{R\}$ of a group G, so that

$$RHR^{-1} = H. \tag{1}$$

The vibrational ground state, the excited state, and the dipole operator each transform like some irreducible representation Γ_m, Γ_n, and Γ_D of the group G. Then one has to form the direct product of any of these irreducible representations, e.g. $\Gamma_m \times \Gamma_D$. If

$$\Gamma_m \times \Gamma_D \in \Gamma_n \tag{2}$$

Table 1. Notations for irreducible representations of point group O_h

M	L	BSW	B	vdLB	HJ
A_{1g}	Γ_1^+	Γ_1	Γ_1^+	α	Γ_s
A_{2u}	Γ_2^-	$\Gamma_{2'}$	Γ_2^-	β	Γ_f^1
A_{2g}	Γ_2^+	Γ_2	Γ_2^+	β'	Γ_i
A_{1u}	Γ_1^-	$\Gamma_{1'}$	Γ_1^-	α'	Γ_k
E_g	Γ_{12}^+	Γ_{12}	Γ_3^+	γ	Γ_d^1
E_u	Γ_{12}^-	$\Gamma_{12'}$	Γ_3^-	γ'	Γ_h
T_{2g}	Γ_{25}^+	$\Gamma_{25'}$	Γ_5^+	ε	Γ_d^2
T_{1u}	Γ_{15}^-	Γ_{15}	Γ_4^-	δ	Γ_p
T_{1g}	Γ_{15}^+	$\Gamma_{15'}$	Γ_4^+	δ'	Γ_g
T_{2u}	Γ_{25}^-	Γ_{25}	Γ_5^-	ε'	Γ_f^2

The notation in the first column is according to Mulliken [12]. The following notations are due to Loudon [8], Bouckaert et al. [13], Bethe [14], von der Lage and Bethe [15], and Howarth and Jones [16]. Indices "g", "u" and $+$, $-$ refer to even and odd parity respectively. Because $O_h = T_d \times I$ this table is also valid for point group T_d when symbols "g", "u", and $+$, $-$ are omitted.

the matrix element may be nonzero, i.e. the transition is allowed by group theory. If

$$\Gamma_m \times \Gamma_D \notin \Gamma_n$$

the matrix element *must* be zero and the transition is forbidden.

For the case of a cubic lattice (point group O_h) the vibrational ground state transforms like A_{1g} while the dipole operator transforms according to T_{1u} (in the following the notation of Mulliken [12] is used; for convenience other notations are listed in Table 1). Therefore, one finds from Eq. (2) that the matrix element

$$(A_{1g}|T_{1u}|\Gamma) \tag{3}$$

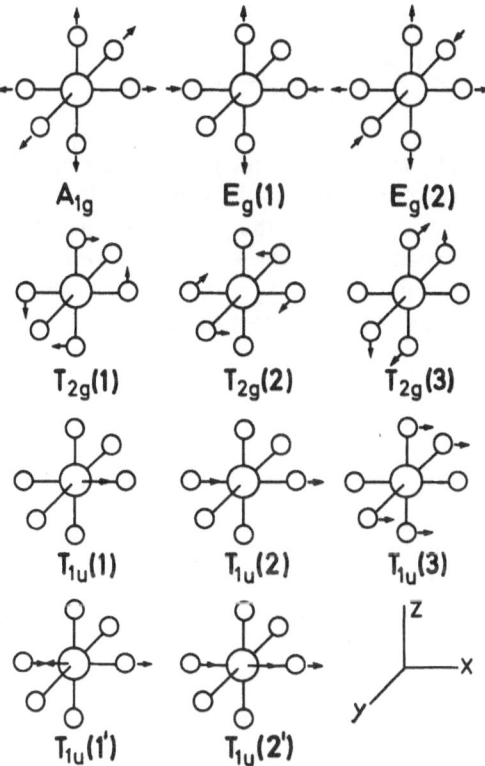

Fig. 2. Illustration of irreducible configurations of an impurity with octahedral symmetry (after Ref. [20]). They correspond to the symmetry vectors of the first nearest neighbor (1 nn) coordinates as introduced in Section 2.4. Configurations A_{1g} and E_g are nondegenerate and twofold degenerate, respectively. Configurations T_{2g} and T_{1u} are threefold degenerate. Linear combinations of $T_{1u}(1)$ and $T_{1u}(2)$ configuration yield configurations $T_{1u}(1')$ and $T_{1u}(2')$. The latter are mainly used in this work. For convenience the prime will, in the future, be omitted

may be non-zero only if $\Gamma \equiv T_{1u}$. The T_{1u} configurations are shown in Fig. 2. Only these will contribute to the one phonon infrared absorption.

While in pure cubic crystals with inversion symmetry (point group O_h) first order Raman scattering is forbidden, in perturbed crystals even-parity oscillations of the nearest neighbors of a defect atom at a site of inversion symmetry may become Raman active. The modes of A_{1g}, E_g, and T_{2g} symmetry contribute to the first order defect induced Raman scattering. This can easily be verified from the conclusions above. The mechanism for scattering has its origin in the modulation of the electronic polarizability by the motion of the impurity and its neighbors. The polarizability can be described by virtual dipolar transitions from low-lying to higher electronic states. Such a process is of third order, and one can write the transition amplitude in the form

$$\sum_{i,j} \frac{(0, A_{1g}|\Gamma_D|i, A_{1g})\,(i, A_{1g}|H_{EP}|j, \Gamma)\,(j, \Gamma|\Gamma_D|0, \Gamma)}{(E - E_{i, A_{1g}}) \cdot (E - E_{j, r})} \tag{4}$$

where 0, and i, j refer to the electronic ground state and the dipole active excited states, respectively. H_{EP} is the electron phonon interaction and transforms like A_{1g}. A_{1g} refers to the phonon ground state while Γ refers to the phonon states excited in the Raman experiment. Γ_D is the dipole operator which transforms like T_{1u}. Then, it is easily seen that Eq. (4) satisfies the condition for non-vanishing matrix elements [Eq. (2)], only if

$$\Gamma_D \times \Gamma_D \in \Gamma.$$

Because

$$T_{1u} \times T_{1u} = A_{1g} + E_g + T_{1g} + T_{2g} \tag{5}$$

the only phonons which can couple to the (virtual) electronic transition are of A_{1g}, E_g, T_{1g}, and T_{2g} symmetry. If the Raman tensor is symmetric

Table 2. First-order selection rules for Raman scattering from point defects in cubic lattices

| | Incident field polarization | |
	[100]	[110]
Scattered field polarization		
$\parallel$	A_{1g}, E_g	A_{1g}, T_{2g}
$\perp$	T_{2g}	E_g

(which is the case for non-resonant Raman scattering) only phonons of symmetry A_{1g}, E_g, and T_{2g} are involved.

The contribution of each symmetry type to the Raman spectrum can be determined from three independent measurements. The types of modes contributing to the spectra for various orientations of incident and scattered electric field polarizations are given in Table 2.

By comparing Eqs. (4) and (5) one finds that in cubic crystals with inversion symmetry (point group O_h) infrared active vibrations are Raman inactive and vice versa. Therefore, infrared absorption and Raman scattering complement each other.

2.3. Localized Modes and Resonant Modes

Ionic crystals have one or more dispersion oscillators (see Section 2.2) whose frequencies are given approximately by the frequencies of the optically active eigenvibrations of the crystal unit cell. Alkali halides which have cubic symmetry (point group O_h) and which have two atoms per unit cell have only one dispersion oscillator, which gives rise to strong optical absorption in the middle to far infrared frequency region. At liquid helium temperature, the crystals are tansparent outside the reststrahl absorption.

The introduction of point defects into a crystal may lead to an activation of optically inactive band vibrations, i.e. to defect induced absorption within the phonon band regions (see Section 2.2 and 4.1), and to eigenvibrations outside the unperturbed band modes. The defect induced band mode absorption is generally very broad and essentially reflects the density of (more or less perturbed) host lattice phonon states weighted by the strength of coupling to the light. In special cases, however, relatively sharp, distinct absorption lines may occur also. This can happen if the defect and/or its nearest neighbors are vibrating at a frequency where the host lattice phonon density is small and/or if this vibration is strongly decoupled from the sourrounding lattice due to strongly reduced local force constants. Such vibrations are called resonant modes or quasilocalized modes. In ionic crystals resonant modes are in general observable only within the acoustic band region, because the optical band is covered by the strong reststrahl absorption.

Eigenvibrations outside the unperturbed band modes of the host lattice are called localized modes. The band width of localized modes is in general very small because direct harmonic damping is impossible. The lifetime of the localized oscillator is determined by indirect coupling to lattice modes. This coupling may be due to anharmonic terms or due to higher order dipole moments in the potential (see Section 3.1.2).

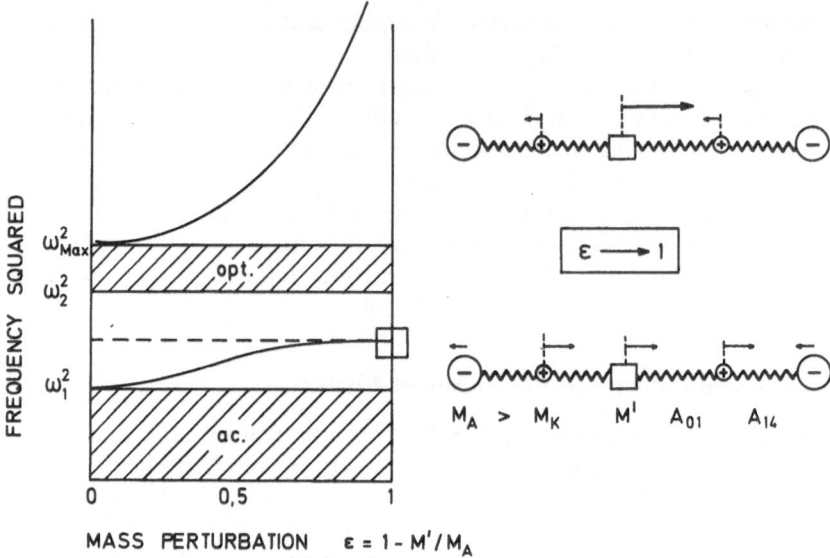

Fig. 3. Localized mode and in-gap mode frequencies as function of substitutional defect mass, M', in the linear diatomic chain approximation with $A_{01} = A_{14} \equiv A$. A refers to the unperturbed force constant of the host lattice. M_K and M_A refer to host lattice cations and anions, respectively. Displacements are shown for the limiting case $M' \ll M_A$, i.e. $\varepsilon = 1 - M'/M_A \rightarrow 1$

In alkali halides localized modes may be located either above the optical band modes or within the acoustic-optical band gap. In the latter case these modes are often called "in-gap modes". The situation can well be demonstrated for the example of defects whose mass, say M', is small when compared with the mass of the host ions, as it is the case e.g. for substitutional hydrogen centers. One gets insight into the problem most easily if one solves a linear diatomic chain model corresponding e.g. to the motion of atoms in the x-direction of Fig. 2 when transverse force constants are neglected. Assuming the defect at the center, one obtains for the case of pure mass perturbation (force constants unchanged) the result of Fig. 3. The main features are two infrared active modes: firstly a high frequency localized mode which splits from the top of the optical band and whose frequency Ω_L approaches infinity for vanishing defect mass, and, secondly, an in-gap mode which splits from the top of the acoustic band and whose frequency stays finite even for $M' = 0$.

The high frequency localized mode is of the configuration T_{1u} (1) in which the defect atom moves out of phase with the first nearest neighbors (see Figs. 2 and 3). This vibration is similar to the reststrahl oscillation where the lattice of positive ions vibrates against the lattice of negative ions. The only difference is that the amplitude is very large at the defect site and decreases exponentially with the distance from the center. It is obvious from this model that the frequency of the $T_{1u}(1)$ vibration strongly depends also on the force constant A_{01} between the defect and first nearest neighbors (1 nn). This force constant may be changed with respect to the host lattice force constant due to a change in the overlap between the defect atom and 1 nn. In fact, this situation can be studied at a $T_{1u}(1)$ vibration which was found for substitutional hydrogen centers in alkali halides. It will be further discussed in Chap. 3.

The in-gap mode has the configuration $T_{1u}(2)$. The displacements of the defect atom and its neighbors are shown for the limiting case $\varepsilon \to 1$ in Fig. 3 (see also Fig. 2). In this vibration the 1 nn move in phase with the defect. The amplitude is not localized at the defect site but is about the same at 1 nn ions. Such a vibration was also found for substitutional hydrogen centers. However, it turned out that the corresponding frequency is not located in the phonon gap but in the upper half of the acoustic band. This frequency shift back into the acoustic band is an effect of the local force constant changes which have to be taken into account. It directly follows from the simple model of Fig. 3 that the mode frequency will sensitively react to force constant changes between 1 nn and fourth nearest neighbors (4 nn). In this model changes in force constants between the defect and 1 nn have no influence at all in the limit $\varepsilon \to 1$, because then the defect and 1 nn vibrate as a "rigid" oscillator against 4 nn. In Chap. 4, $T_{1u}(2)$ type in-gap modes and resonant modes associated with color centers as mentioned in Section 2.1 will be discussed. Examples of infrared active in-gap modes and resonant modes due to defects in polar and homopolar crystals in general are given in reviews by Genzel [17], and by Sievers [18].

The introduction of defects into a crystal can lead also to defect induced Raman scattering from (somewhat perturbed) host lattice band modes, and to Raman active localized modes and resonant modes. When the defect atom is at a site of inversion symmetry only the even-parity oscillations of the nearest neighbors are Raman active (see Section 2.2). This case was studied in connection with F centers in alkali halides. It will be discussed in Chap. 5. When the defect atom is *not* at a site of inversion symmetry, the motion of the impurity itself may be Raman active in a localized or a resonant mode. The incident (monochromatic) light of frequency ω_i is then scattered to frequencies $\omega_i \pm \omega_L$ or $\omega_i \pm \omega_R$, respectively. This case was not yet observed for centers discussed in this work, however, it could eventually occur for interstitial hydrogen centers which are discussed in Section 3.2.

2.4. Theoretical Background

In this section only a few remarks on the theory of infrared absorption
and Raman scattering of crystal lattices with point defects are made.

The theory of infrared absorption of crystal lattices with point defects
was reviewed by several authors as e.g. by Benedek and Nardelli [19],
and by Klein [20]. The linear absorption coefficient is, neglecting reflection, defined in terms of the intensity of the transmitted light relative to
the incident light, I/I_0, and is given by

$$\alpha = (1/d) \ln(I/I_0) = (4\pi\omega/cn(\omega)) \operatorname{Im}\chi(\omega) \tag{6}$$

where d is the sample thickness, ω the frequency of incident light, c is the
velocity of light and $n(\omega)$ is the refractive index. The dielectric sus-
ceptibility χ is a tensor in general. For cubic crystals one has $\chi_{\alpha\beta} = \chi \cdot \delta_{\alpha\beta}$.
The susceptibility is originally obtained from a linear response theory.
Using the Green function formalism one can write

$$\chi = (NZ^2/v_0\mu) \lim_{\varepsilon \to 0} (\varphi_{T0}|[G_0^{-1} + pG_0^{-1}TG_0]^{-1}|\varphi_{T0})$$

$$= (NZ^2/v_0\mu) \lim_{\varepsilon \to 0} [(\omega_{T0}^2 - (\omega + i\varepsilon)^2)^2 + p(\varphi_{T0}|T|\varphi_{T0})]^{-1} \tag{7}$$

where N is the number of unit cells, and v_0 the cell volume. Z is the effective
charge, μ the reduced mass, and p the probability of finding one defect
in the unit cell. The T matrix is given by

$$T = V(I + G_0V)^{-1}$$

where G_0 is the Green function for the unperturbed lattice. V is the
perturbation matrix which contains the perturbation in mass and force
constants. V and T are generally defined in a $3rN$ dimensional vector
space (r is the number of particles in the unit cell). However, if the perturbed region within the lattice is sufficiently localized, the elements of
V and T are non-zero only in a $3q$-dimensional subspace, the so called
"impurity space" (q is the number of ions affected by the perturbation).
ω_{T0} is the frequency of the dispersion oscillator with $k \approx 0$. φ_{T0} is the
corresponding eigenvector.

To simplify the calculation of the T-matrix elements one can introduce
special symmetry vectors (shell vectors), which are described in more
detail in Refs. [21] and [22]. These symmetry vectors constitute a
complete set of basis functions and are most suitable for describing the
lattice vibrations around a defect with a finite range of perturbation.

In a lattice with point-symmetry group G the lattice ions may be
assumed to be arranged on "shells" around the defect. A "shell" is defined
by all lattice points which transform into one another under all symmetry

operations of G. Symmetry vectors are those basis vectors to irreducible representations of G, the amplitudes of which are nonzero only on a single shell. The symmetry vectors are described by four indices. s is the shell number, starting with $s = 0$ for the defect itself, Γ gives the irreducible representation of G, r is the index of multiplicity of Γ in the shell s, and j is the index of degeneracy. For the projections $s(\Gamma, j, s, r)$ of the eigenvectors φ_{T0} one obtains

$$s(\Gamma, j, s, r) \equiv (\varphi_{T0} | \sigma(\Gamma, j, s, r))$$

and in analogy

$$\hat{T} = (\sigma | T | \sigma') = (\sigma | V(I + G_0 V)^{-1} | \sigma')$$

$$= \sum_{\sigma''} (\sigma | V | \sigma'')(\sigma'' | R | \sigma') \equiv \hat{V}\hat{R} \, . \tag{8}$$

For calculating χ one can substitute the second term in Eq. (7) by $(s | \hat{T} | s)$.

Assuming that the translational symmetry of the effective charges is *not* disturbed in the perturbed lattice, one can write the matrix of effective charges in the form

$$Z = I \times Z_r, \qquad \text{where } Z_r = e\begin{pmatrix} +1 \\ & -1 \end{pmatrix}.$$

As a consequence only those infrared-active modes which can be deduced from T_{1u} modes of the unperturbed lattice will contribute to the absorption coefficient in the perturbed lattice, i.e., $s \neq 0$ only if $\sigma = \sigma(T_{1u}, j, s, r)$. Hence

$$(s| = (s(T_{1u}), 0| \, .$$

This results in a further simplification of the projected perturbation matrix $\hat{V}$ and the Green function $\hat{G}$. The scalar product $(s | \hat{T} | s)$ has the simple form

$$(s | \hat{T} | s) = (s(T_{1u}) | \hat{V}(T_{1u}) [I + \hat{G}(T_{1u}) \hat{V}(T_{1u})]^{-1} | s(T_{1u})) \, . \tag{9}$$

To find the only unknown in Eq. (9) one has to discuss the vector $(s(T_{1u})|$ and the projected perturbation matrix $\hat{V}(T_{1u})$ for special models. This will be done only in Section 4.2.3 for the case of infrared absorption of F centers. For other cases the reader is refered to the original literature.

The specific force constant model to be used in the following sections is shown in Fig. 4. Changes in force constants between the defect and 1 nn, $A01 = (A_{01} - A)/A$, and between 1 nn and 4 nn, $A14 = (A_{14} - A)/A$, are taken into account. If $\Delta k = 0$ this model corresponds to the extended model by Gethins et al. [23].

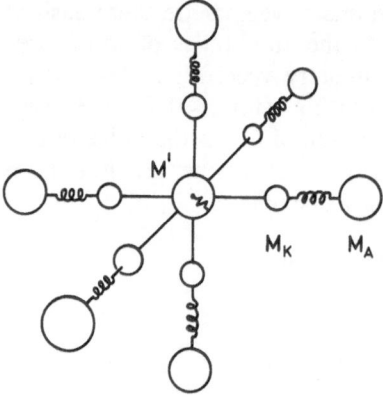

Fig. 4. Force constant model essentially discussed in this work. —— A_{01} is the force constant between the defect and 1 nn, ∿∿ A_{14} between 1 nn and 4 nn. ∿∿∿ k is the shell-core spring

The theory of Raman scattering from crystals with point defects was developed by Sennett [24], Xinh [25] Maradudin [26], Benedek and Nardelli [27], and some others.

The relation between the intensities of the incident light of frequency ω_i and the scattered light of frequency ω_s can be described by the fourth-order tensor

$$i_{\alpha\gamma\beta\lambda}(\Omega) = (2\pi)^{-1} \int dt \, e^{-i\Omega t} \langle P_{\beta\lambda}(t) P_{\alpha\gamma}^*(0)\rangle \tag{10}$$

where $\Omega = \omega_i - \omega_s$ is the energy transfer, and $P_{\beta\lambda}(t)$ is the time dependent operator for the electronic polarizability tensor of the crystal. One now can expand the polarizability in series of lattice ion displacements, say $u(l, k)$ (l being a Bravais vector, and k a cell index). The zero order term (no phonons involved) accounts for Rayleigh scattering ($\Omega = 0$), while the first and higher order terms account for first and higher order Raman scattering, respectively. The first order process is characterized by the coefficient.

$$P_{\alpha\beta,\mu} \equiv \partial P_{\alpha\beta}/\partial u_\mu(l, k).$$

As it was shown by Maradudin [26] the first order part of Eq. (10) can be written for the Stokes process in the form

$$i_{\alpha\gamma\beta\lambda}(\Omega) = (p\hbar/\pi M)(n(\Omega) + 1) \sum_{\substack{l, k, \mu \\ l', k', \mu'}} P_{\beta\lambda,\mu}(l, k) \, P_{\alpha\gamma,\mu'}(l', k')$$

$$\times \operatorname{Im}(l, k, \mu | [I + G_0 V]^{-1} G_0 | l', k', \mu') \tag{11}$$

where the same notations as above have been used. M is the mass of 1 nn of the defect. In analogy to the treatment given above one now can introduce symmetry vectors in order to simplify the calculation of Eq. (11).

For cubic symmetry there are only three independent components of the Raman tensor

$$i_{zzzz} = i_{11} \; ; \; i_{xxyy} = i_{12} \; ; \; i_{xyxy} = i_{44} \; .$$

For a perturbation extending only to 1 nn of the defect, there are three different non-zero terms

$$P_{xx,xx} \equiv \tfrac{1}{3}(\alpha + 2\beta); \; P_{xx,yy} \equiv \tfrac{1}{3}(\alpha - \beta); \; P_{xy,xy} \equiv \gamma \; .$$

Then the components of the first order Raman tensor for cubic crystals are

$$i_{11} = (p\hbar/6\pi M \Omega)(n(\Omega) + 1)\left[\alpha^2 \hat{\varrho}(A_{1g}) + 2\beta^2 \hat{\varrho}(E_g)\right]$$

$$i_{12} = (p\hbar/6\pi M \Omega)(n(\Omega) + 1)\left[\alpha^2 \hat{\varrho}(A_{1g}) - \beta^2 \hat{\varrho}(E_g)\right] \qquad (12)$$

$$i_{44} = (p\hbar/2\pi M \Omega)(n(\Omega) + 1)\left|\gamma^2 \hat{\varrho}(T_{2g})\right|$$

where $\hat{\varrho}(\Gamma)$ is the projected one-phonon density

$$\hat{\varrho}(\Gamma) = 4M\Omega \operatorname{Im}(\sigma|\,[I + G_0 V]^{-1}\, G_0|\sigma') \; .$$

The coefficients α, β, γ can be obtained experimentally from uniaxial stress experiments. For the case of F centers they were determined by Gebhardt and Maier [28] for various alkali halides from the stress splitting of the electronic F band.

3. Infrared Vibrational Absorption: High Frequency Localized Modes

High frequency localized modes occur well above the maximum host lattice frequency. Such modes were found for substitutional hydrogen ions, $U(H_s^-, D_s^-)$ centers, in a number of compounds. Extensive experimental and theoretical investigations were performed in alkali halides ([30–49], and [22, 23, 29, 40, 50–64] respectively), and in alkaline earth fluorides ([5, 65–67], and [5] respectively). The results will be reviewed in Section 3.1. High frequency localized modes due to interstitial hydrogen ions, $U_1(H_i^-, D_i^-)$ centers, were studied experimentally in a number of alkali halides [69–72], and in CaF_2 [7]. The only theoretical investigation was performed for the case of $KBr : H_i^-$ [61]. Section 3.2 reviews the results. High frequency localized mode absorption due to interstitial hydrogen atoms in LiF [73, 74], and in CaF_2 [75] is fairly well established. The results are mentioned in Section 3.3.

With respect to electron centers like F centers, Rosenstock and Klick [29] have argued that the frequency of the electronic transition $1s$–$2p$ could be explained by a classical "mechanical" vibration of the electron. The model used by these authors is linear, and does not include local force constant changes. For KCl they obtained for the "F center localized mode frequency" about 3.2 eV, while the measured value of the $1s$–$2p$ transition is about 2.3 eV at 4.2 K. A much better approximation can be obtained when calculating the F center localized mode frequency from the experimentally determined main line frequency for the H_s^- center localized mode by taking into account only the mass perturbation, i.e. by assuming that the local force constants, in particular A_{01}, are the same for the electron and the hydrogen ion (see also Section 4.2.4). One then obtains values which reproduce the frequency of the visible F band within 5–20% (for KCl one estimates about 2.4 eV). However, such a mechanical model of the electronic transition is not in particular useful.

The investigations reported in this chapter should serve to give an idea about the information which can be obtained from optical investigations of defective crystal lattices. In particular, it is shown how local site symmetries of hydrogen centers in various compounds can be derived from the number of vibrational transitions observed. The frequency positions and intensities of lines allow conclusions on the local oscillator potential, and on local vibrational amplitudes respectively. The investigation of temperature dependent line widths and frequency shifts, and of additional absorption structures observed, yields information on the nature and the amount of phonon-phonon coupling, and on host lattice phonon densities. The results may apply to pure host lattices as well, i.e. the structure and damping of host lattice dispersion oscillators may be discussed along similar lines.

3.1. U Centers (H_s^-, D_s^- Centers)

3.1.1. The Main Line Frequency

Fig. 5 shows a typical localized mode absorption spectrum for the example of H_s^- centers in KBr. This absorption is located in the near infrared spectral region. At room temperature the absorption shape is broad and structureless. Sidebands, which are located nearly symmetrically to the main line at 445 cm^{-1} are observed with decreasing temperature. The main line is due to the motion of the hydrogen ion in a cubic

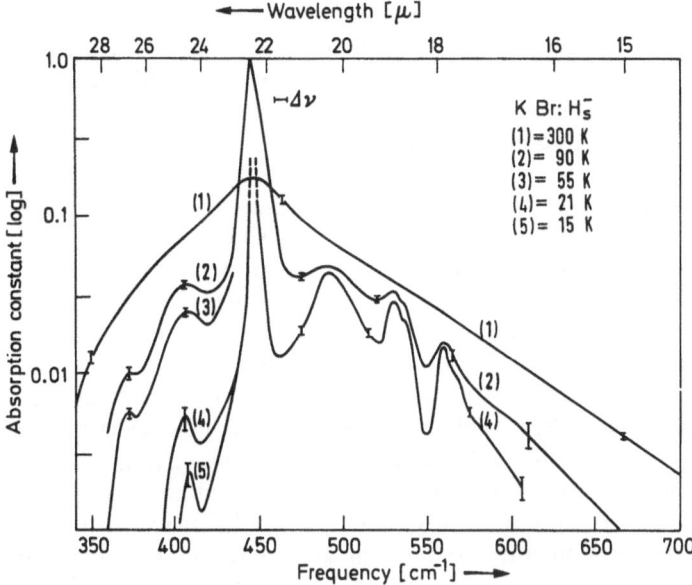

Fig. 5. Infrared absorption of the H_s^- center localized mode in KBr (after Ref. [37])

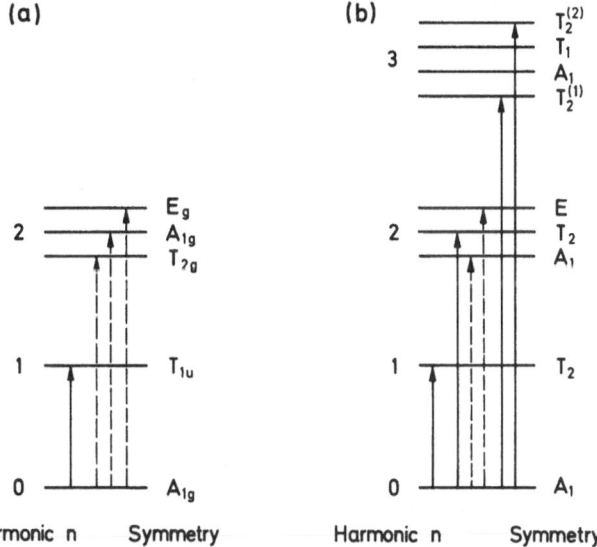

Fig. 6. (a) Energy level scheme of an anharmonic oscillator in a cubic (O_h) environment corresponding to U centers in alkali halides. Full arrow corresponds to an infrared active transition. Transitions indicated by dotted arrows are only Raman active. (b) Same as (a) but for T_d symmetry. Transitions observed in the infrared are shown by full arrows. Dotted arrows indicate infrared transitions which are induced by shear strain (after Ref. [82])

potential. The line is threefold degenerate and corresponds to a vibrational transition from an A_{1g} ground state $(n=0)$ to a $T_{1u}(1)$ excited state $(n=1)$. This transition is shown in Fig. 6 (a). The transition energy is listed in Table 3 for various alkali halides. A model for the $T_{1u}(1)$ vibration is shown in Fig. 2 of Section 2.2

For a phenomenological description of the main line frequency one can treat the hydrogen ion as moving in a *static* potential of O_h symmetry. This approach is physically reasonable since, during one period of oscillation of the "heavy" neighbor ions in a (perturbed) host crystal band mode, the "light" hydrogen ion makes many complete oscillations, so that during any of them, it sees the surrounding ions as essentially at rest. This strong localization of the U center vibration may be directly seen from the isotope shift in the frequency which is about $\sqrt{2}$ for H_s^- and D_s^- centers (see Table 3). Such a value is expected from the mass ratio for "harmonic" H_s^- and D_s^- oscillators moving in a rigid box. In this approximation 1 nn of the hydrogen are held fixed (compare Fig. 4).

A further support of this simple model follows from the localized mode absorption intensities which are related to the vibrational displacements. One expects the integrated absorption for the H_s^- localized

Table 3. Main line frequency and halfwidth for U center localized mode in alkali halides, and alkaline earth fluorides[a]

Substance	Frequency $\nu_L(H_s^-)$ [cm^{-1}]	Halfwidth $\Delta\nu_L(H_s^-)$ [cm^{-1}]	Frequency ratio $\nu_L(H_s^-)/\nu_L(D_s^-)$	Temperature [K]	References
[6]LiF	1030.9			20	[18]
LiF	1027	4.2	1.38	20	[38]
	1015	18	1.37	300	[44]
NaF	859.5		1.40	20	[38]
	846.7	12	1.39	300	[44]
NaCl	563	5.6	1.38	90	[30, 37]
	565	43		300	[37]
NaBr	498	17	1.38	90	[30, 37]
	504	60		300	[30, 37]
NaI	426.8		1.34	10	[41]
	430.6			100	[31]
KF	725.5			100	[31]
KCl	502	2.3	1.40	90	[30, 32, 33, 37]
	497	32	1.38	300	[37]
KBr	446	6	1.40	90	[30, 32, 33]
	444			300	[37, 40]
	886.3 (S, T_{2g})		1.40	11	[76] R
	894.8 (S, A_{1g})			11	[76] R
	904.5 (S, E_g)		1.40	11	[76] R

Table 3 (continued)

Substance	Frequency $v_L(H_s^-)$ [cm^{-1}]	Halfwidth $\Delta v_L(H_s^-)$ [cm^{-1}]	Frequency ratio $v_L(H_s^-)/v_L(D_s^-)$	Temperature [K]	References
KI	382	14.5	1.37	90	[30, 31, 37]
	392			300	
	753.1 (S, T_{2g})		1.39	8	[76] R
	759.9 (S, T_{2g})			8	[76] R
	767.9 (S, E_g)		1.40	8	[76] R
RbF	703.1			100	[31]
RbCl	476	4.8	1.40	90	[30, 37]
	466		1.38	300	[30, 37]
RbBr	425	8		90	[30, 37]
	412			300	[30, 37]
RbI	360			70	[30]
CsCl	425		1.40	20	[45]
	417			300	[45]
CsBr	364.1	0.4	1.42	6	[36, 49]
	363	35	1.40	300	[36, 49]
CsI	282.8	0.6	1.41	7	[45, 49]
	303		1.38	300	[45, 49]
CaF$_2$	965.6	<0.7	1.39	20	[5]
	969	5.0		16	[77] R
	957.8	8.7	1.39	290	[5]
	1895 (S, A_1)	<4.0		16	[77] R
	1894.0 ,,			77	[78] SI
	1919.8 (S, T_2)		1.39	20	[5]
	1923 ,,	<4.0		16	[77] R
	1945.0 (S, E)			77	[78] SI
	2912.2 (T, T_2)		1.39	20	[5]
	2825.6 (T, T_2)		1.38	20	[5]
SrF$_2$	893.2	1.7	1.39	20	[5]
	896	4.8	1.40	16	[77] R
	1747 (S, A_1)	<.4	1.39	16	[77] R
	1775.9 (S, T_2)		1.39	20	[5]
	1778 ,,	<4	1.39	16	[77] R
BaF$_2$	806.6	8.8		20	[5]
	805	2.2	1.40	16	[77] R
	798.2	14		290	[5]
	1566 (S, A_1)	<4.2	1.38	16	[77] R
	1596.2 (T, T_2)			20	[5]
	1597 ,,	<4.2	1.39	16	[77] R

[a] If not indicated further, values refer to infrared active fundamentals. Higher harmonic transitions were included in the Table for convenience. Their values follow parantheses. First symbol in parantheses, i.e. S or T, refers to second harmonic ($n = 2$), and third harmonic ($n = 2$) frequencies respectively. Second symbol in parantheses refers to the symmetry of the excited vibrational state. R after reference means measured by Raman scattering; SI after reference means stress induced transition.

mode to be twice that of the D_s^- mode. Experimentally a ratio of 1.92 is found [33, 43].

Calculating the main line frequency on the basis of the simple model discussed, and assuming *pure* mass perturbation (setting the force constant A_{01} equal to the host lattice force constant A) one obtains values which are about 40% to 60% too high when compared with experimental values as listed in Table 3. This indicates that a considerable lowering in local force constants has to be taken into account. In fact, this was confirmed in more sophisticated calculations using Green function techniques. On the other hand, these calculations did show also that for a description of further details of the localized mode absorption, force constant changes at least up to 4 nn have to be taken into account. It is returned to this point in Section 3.1.4. In spite of the usefulness of the simple model described above, a rigorous treatment of the U center vibrational problem requires the dropping of both assumptions: the complete localization, and the harmonic motion of the hydrogen oscillator. The motion of the neighboring ions will influence the main line frequency for H_s^- and D_s^- centers differently. This is due to the different degree of localization for H_s^- and D_s^- vibrations (a measure of the degree of localization is, roughly speaking, the distance of the main line frequency from the maximum host lattice frequency). In fact, the deviation from the value $\sqrt{2}$, observed for the frequency ratio for H_s^- and D_s^- vibrations, may be explained in part by including the motion of neighboring ions. On the other hand, the "large" deviation observed in this frequency ratio, cannot be explained in the harmonic approximation alone. Note that for NaI : H_s^-, D_s^- one finds, e.g., $v_L(H_s^-)/v_L(D_s^-) = 1.34$! This indicates that fairly large anharmonic contributions to the harmonic localized mode frequency may be present as well.

In the following the simple model discussed above will be extended by including anharmonic terms in the oscillator potential, which is still assumed to be static. Expanding the potential energy in a power series in the displacements of the hydrogen ion from equilibrium one can write the Hamiltonian in the form

$$H = \tfrac{1}{2}M'\Omega_L^2 r^2 + C_1(x^4 + y^4 + z^4) + C_2(x^2 y^2 + x^2 z^2 + y^2 z^2) + \ldots \quad (13)$$

where M' is the hydrogen mass and $r^2 = x^2 + y^2 + z^2$. Because the anharmonic terms are expected to be small when compared with the harmonic term the unperturbed wave functions are those of a spherically symmetric harmonic oscillator, having energy levels

$$E_n = \hbar\Omega_L(n + \tfrac{3}{2}). \quad (14)$$

These equally spaced degenerate levels are split and shifted by the anharmonicity (see Fig. 6). In O_h symmetry, where cubic terms in Eq. (13) are absent, the first excited state $n = 1$ is only shifted. Because of the different vibrational amplitude for H_s^- and D_s^- centers, this anharmonic shift is of different size for H_s^- and D_s^- levels. It explains in part the deviation from the harmonic frequency ratio. The sixfold degenerate $n = 2$ level is split by the anharmonicity into levels of A_{1g}, E_g, and T_{2g} symmetry. The corresponding transitions from the A_{1g} ground state $(n = 0)$ to the $n = 2$ excited states are "only" Raman active and will be discussed in Section 5.1. The quartic terms in Eq. (13) allow two infrared active transitions from the ground state to excited states with $n = 3$. For H_s^-, D_s^- centers in alkali halides these transitions were not yet found. Recently, Akhvlediani and Politov [73] correlated a line observed at $1900\ \mathrm{cm}^{-1}$ in neutron irradiated LiF : OH$^-$ crystals with a third harmonic transition of T_s^- (tritium) centers. This correlation, however, seems somewhat arbitrary; furthermore, the high oscillator strength of this transition is not understood from own experiments with H_s^- and D_s^- centers. In particular, Dötsch [80] could not observe a third harmonic transition in LiF and NaF even with H^- center concentrations up to 1 Mol%. However, before discussing the influence of anharmonicity further, we shall turn briefly to the results in alkaline earth fluorides and rare-earth trifluorides.

In alkaline earth fluorides the U center has tetrahedral site symmetry (T_d). One therefore has to add to the expansion of Eq. (13) a term of the form $Bxyz$. As a result, in the energy level scheme shown in Fig. 6 (b) a transition between the ground state $(n = 0)$ and the second excited state $(n = 2)$ (full arrow) becomes infrared active[1]. Second and third harmonic transitions were observed in $CaF_2 : H_s^-, D_s^-$ in $SrF_2 : H_s^-, D_s^-$ and in $BaF_2 : H_s^-$ by Elliott et al. [5]. The corresponding frequencies are listed in Table 3. Transition frequencies obtained from Raman scattering [77], and from experiments with uniaxial stress [78] are included in the Table for convenience. The linewidths should be viewed with some caution, since they are markedly concentration dependent for the high defect center concentrations used. It should be mentioned that experiments with tunable CO_2 lasers did enable one to study the step-wise excitation for each of the three $n = 2$ levels [81].

The measured transition frequencies can be used to calculate the anharmonic coefficients Ω, B, C_1, and C_2. In this connection Elliott et al. [5] did estimate the effect of anharmonic terms in the Hamiltonian by perturbation theory. The energy for the perturbed vibrational levels relative to the unperturbed ground state can be written in the form

$$E_n(\Gamma) = n\hbar\Omega_L + (\hbar/(2M'\Omega_L))^2 (C_1\mu_1(\Gamma) + C_2\mu_2(\Gamma))$$
$$- B^2\lambda(\Gamma)\hbar^2/(24M'^3\Omega_L^4) \tag{15}$$

---

[1] One could, on the other hand, argue that the observation of a second harmonic indicates that in alkaline earth fluorides the local site symmetry of the hydrogen ion is T_d.

where μ_i and λ are tabulated for representations Γ under consideration in Table 4. For the case of O_h symmetry one has to set $B \equiv 0$. Values for Ω, B, C_1, and C_2 as obtained from Eq. (15) and the measured transition frequencies of Table 3 are listed in Table 5. The corresponding values for U centers in KBr and KI are included. They are based on Raman scattering experiments (see Chap. 5). With the anharmonic coefficients of Table 5 one can calculate from Eq. (15) the complete energy level schemes of Fig. 6.

The existence of localized modes due to hydrogen and deuterium ions in the rare-earth trifluorides LaF_3, CeF_3, PrF_3, and NdF_3, was reported by Jones and Satten [79]. The infrared spectra show two strong absorption lines [for LaF_3 e.g. at $818\,cm^{-1}$ and at $1168\,cm^{-1}$ (4.2 K)], which are probably due to hydrogen ions on fluorine sites with D_3 symmetry. The substitution of deuterium results in frequencies which are reduced by a factor of 1.4. These modes were also established from the

Table 4. Values $\mu_i(\Gamma)$ and $\lambda(\Gamma)$ for anharmonic localized oscillator in cubic environment

Harmonic n	Symmetry Γ	$\mu_1(\Gamma)$	$\mu_2(\Gamma)$	$\lambda(\Gamma)$
0	A_1	9	3	1
1	T_2	21	7	5
2	A_1	45	15	21
	E	45	9	3
	T_2	33	15	13
3	A_1	45	27	25
	T_1	57	21	15
	$T_2^{(1)}$	$\begin{pmatrix} 57 & 0 \\ 0 & 81 \end{pmatrix}$	$\begin{pmatrix} 25 & 2\sqrt{6} \\ 2\sqrt{6} & 15 \end{pmatrix}$	$\begin{pmatrix} 27 & 6\sqrt{6} \\ 6\sqrt{6} & 13 \end{pmatrix}$
	$T_2^{(2)}$			

Table 5. Coefficients for anharmonic oscillator potential for U centers in alkali halides and alkaline earth fluorides

Substance	$\Omega_L(H_s^-)$ [cm^{-1}]	$\Omega_L(D_s^-)$ [cm^{-1}]	B $\times 10^{12}$ [erg/cm^3]	C_1 $\times 10^{19}$ [erg/cm^4]	C_2 $\times 10^{19}$ [erg/cm^4]	Temperature [K]	References
KBr	444.6	318.2	0	0.986	-2.30	10	[76] R
KI	377.8	272.3	0	0.576	-1.356	8	[76] R
CaF_2	981.1	702.1	7.87	-1.01	-2.32	20	[82]
SrF_2	907.4	[a]	6.20	-2.19	-1.80	20	[78]
BaF_2	827.2		3.98	-8.5	-1.7	20	[78]
	817		3.36	-1.35	3.06	4.2	[77]

[a] Approximately one can set $\Omega_L(D_s^-) \approx \Omega_L(H_s^-)/\sqrt{2}$.

vibronic spectra. In addition to the two intense lines, four absorption lines with smaller intensity were observed. These additional lines may be due to hydrogen ions on other sites which are occupied to a lesser extent. A strong correlation of these lines is not yet possible. One difficulty is that the crystal structure of the rare-earth trifluorides is not completely clarified. If the structure belongs to the space group D_{3d}^4, there are three fluorine sites which have D_3, C_3, and C_1 symmetry, respectively.

3.1.2. Multiphonon Processes

The coupling of the U center localized mode to lattice modes becomes evident in the temperature dependent halfwidth and frequency shift of the main line, and in the occurence of phonon sidebands. It was shown by Bilz et al. [51] and by Elliott et al. [5] that for the case of U centers in alkali halides and alkaline earth fluorides this coupling is mainly of anharmonic nature. Contributions due to higher-order dipole moments are small. For a complete description of the observed anharmonic effects further terms which contain combinations of H_s^- ion displacements and displacements of ions on nearby lattice sites must be added to Eq. (13). Expanding the potential energy again up to terms of fourth order and transforming to phonon creation and annihilation operators, one can write the Hamiltonian for the localized oscillator in the form

$$H = H_0 + H_A \tag{16}$$

where H_0 is the harmonic part of the Hamiltonian. The anharmonic part can be written in the form

$$
\begin{aligned}
H_A = & \frac{1}{3!} \sum_{L,i,j} V_3(L,i,j)(a_L + a_L^+)(a_i + a_i^+)(a_j + a_j^+) \\
& + \frac{1}{4!} \sum_{L,i,j,k} V_4(L,i,j,k)(a_L + a_L^+)(a_i + a_i^+)(a_j + a_j^+)(a_k + a_k^+) + \dots
\end{aligned}
\tag{17}
$$

Indices i, j, k refer to lattice phonons. The influence of H_A on the localized mode can be made plausible by introducing diagrams as shown in Fig. 7. These symbolize transitions between quantum states. Straight and bent lines are introduced for illustrating phonon-phonon (anharmonic) interactions. Single lines refer to lattice phonons. Double lines refer to the localized oscillator. At vertices (intersection points) one has to connect a number of lines corresponding to the order of anharmonicity (three for third order anharmonicity via V_3, and four for fourth order anharmonicity via V_4). Closed diagrams with *one* vertex yield a contribution to the self energy in *first* order, diagrams with n vertices in n th

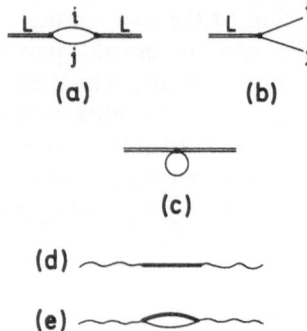

Fig. 7. Self energy diagrams illustrating phonon-phonon (anharmonic) interactions [cases (a−c)], and photon-phonon interactions [cases (d), (e)]. Double lines refer to the localized oscillator (Ω_L), single lines to lattice phonons ($\omega_i, \omega_{j...}$). (a) Localized oscillator frequency shift $\Delta_L(\omega)$ due to two phonon decay into virtual intermediate states through $V_3^2(L, i, j)$. (b) Imaginary part of self energy, characterizing damping due to two phonon decay. This diagram is obtained by bisecting diagram (a). Interaction through $V_3^2(L, i, j)$. (c) Frequency shift through coupling to phonon ω_i. Interaction through $V_4(L, L, i, j)$. − Diagrams (a)-(c) may follow processes (d) and (e). (d) Resonance (undamped) between photon and localized oscillator Ω_L. Interaction through first order dipole moment M_1. (e) Same as case (d) but for $\omega = \Omega_L \pm \omega_i$. Interaction through second order dipole moment M_2. This process is negligible for U centers in alkali halides and alkaline earth fluorides

order. For example the process in Fig. 7 (a) is associated with V_3^2, and (c) with V_4. Oscillating lines illustrate photons. They describe in connection with straight or bent lines photon-phonon interaction.

Fig. 7 (a) shows a typical third order contribution to H_A: An intermediate state with two excited phonons is coupled by $V_3^2(L, i, j)$ to the localized mode, resulting in a shift of the harmonic oscillator frequency Ω_L. Diagram (b) is obtained by bisecting diagram (a). It represents the imaginary part of the self energy, and describes the damping due to decay into two arbitrary band phonons ω_i and ω_j. All contributions of this type are obtained by summing over all ω_i and ω_j within the phonon bands. Of course, energy must be conserved in these processes, i.e. $\Omega_L = \omega_i \pm \omega_j$. A different type of a closed diagram which is associated with fourth order anharmonicity is shown in Fig. 7 (c). In this case the localized oscillator interacts virtually with one phonon. This process is independent of frequency. It therefore contributes to the shift but not to the damping.

Diagrams as introduced in Fig. 7 can be used for a systematic description of all contributions to the complex susceptibility. They can be built up from the simple diagrams (a)-(c) in Fig. 7. A survey on the most important contributions in connection with the U center localized mode is given in Table 6. A more complicated process which is discussed by

Table 6. Contributions to anharmonic frequency shift and damping. (s) refers to sum processes and (d) refers to difference processes

	Frequency shift $\Delta_L(\omega)$	Damping $\Gamma_L(\omega)$	$\Delta_L(\omega)$	$\Gamma_L(\omega)$	Temperature dependence $\Delta_L(\omega), \Gamma_L(\omega)$
1.		high frequency sideband (s) low frequency ~ (d)	$-\dfrac{2(\Omega_L \pm \omega_i)}{(\Omega_L \pm \omega_i)^2 - \omega^2}$	$\delta(\Omega_L \pm \omega_i - \omega)$	$n_L + n_i + 1 \approx n_i + 1$ (s) $-n_L + n_i \approx n_i$ (d)
2.		2-phonon sidebands (s)	$-\dfrac{2(\Omega_L + \omega_i \pm \omega_j)}{(\Omega_L + \omega_i \pm \omega_j)^2 - \omega^2}$	$\delta(\Omega_L + \omega_i + \omega_j - \omega)$	$(n_L + 1)(n_i + 1)(n_j + 1)$ $-n_L n_i n_j \approx (n_i + 1)(n_j + 1)$ (s)
3.		halfwidth (H_s^- and D_s^-) halfwidth (D_s^-)	$-\dfrac{2(\omega_i \pm \omega_j)}{(\omega_i \pm \omega_j)^2 - \Omega_L^2}$	$\delta(\omega_i - \omega_j)$ $\delta(\omega_i \pm \omega_j - \Omega_L)$	$3[(n_L + 1)(n_i + 1)n_j$ $-n_L n_i(n_j + 1)] \approx 3(n_i + 1)n_j$ (d) $n_i + n_j + 1$ (s)
4.		halfwidth (H_s^- and D_s^-)	$-\dfrac{2(\omega_i + \omega_j \pm \omega_k)}{(\omega_i + \omega_j \pm \omega_k)^2 - \Omega_L^2}$	$\delta(\omega_i + \omega_j \pm \omega_k - \Omega_L)$	$(n_i + 1)(n_j + 1)(n_k + 1)$ $-n_i n_j n_k$ (s)
5.		halfwidth (H_s^- and D_s^-)	$-\dfrac{2(\Omega_L + \omega_j - \omega_k)}{(\Omega_L + \omega_j - \omega_k)^2 - \Omega_L^2}$	$\delta(\omega_j - \omega_k)$	$3[(n_L + 1)(n_j + 1)n_k$ $-n_L n_j(n_k + 1)] \approx 3(n_j + 1)n_k$ (d)
6.		—	$-1/\omega_j$	0	$2n_i + 1$
7.		—	$+1$	0	$2n_i + 1$

Elliott et al. [5] is shown in the fifth row of Table 6. A complete survey on the anharmonic processes is given by Bilz et al. [51].

Because the term H_A in Eq. (16) is small when compared with H_0, the contributions of multiphonon processes to the energy and damping of the localized oscillator can be calculated by perturbation theory. For the diagram in the third row of Table 6 the first term of Eq. (17) for H_A yields, in lowest order, the linewidth[2]

$$2\Gamma(\omega_L) = 36\pi \sum_{i,j} |V_3(L,i,j)|^2 \left[(\bar{n}_i + 1)(\bar{n}_j + 1) - \bar{n}_i \bar{n}_j\right] \delta(\Omega_L - \omega_i - \omega_j) \quad (18)$$

and the frequency shift

$$\Delta(\omega_L) = 36 \sum_{i,j} \frac{|V_3(L,i,j)|^2 (\omega_i + \omega_j)}{\Omega_L^2 - (\omega_i + \omega_j)^2} \left[(\bar{n}_i + 1)(\bar{n}_j + 1) - \bar{n}_i \bar{n}_j\right]. \quad (19)$$

Here difference processes for phonons ω_i and ω_j are not included. Further results for the shift function $\Delta_L(\omega)$ and the damping function $\Gamma_L(\omega)$ are listed in the third and fourth column of Table 6. Only constant factors and the anharmonic coefficients $|V_n(L, i \dots)|^l$ have been omitted. It is understood that $\Delta_L(\omega)$ and $\Gamma_L(\omega)$ are not independent of one another. They are real and imaginary parts of a damping function

$$\Pi_L(\omega) = \Delta_L(\omega) + i\Gamma_L(\omega)$$

and are related by the Kramers-Kronig-relations.

The temperature dependence of $\Delta_L(\omega)$ and $\Gamma_L(\omega)$ is listed in the fifth and sixth column of Table 6. It can be directly calculated by using the properties of boson operators a^+ and a.

$$\begin{aligned} a_i^+ \ |n_i) &= (n_i + 1)^{1/2} |n_i + 1) \\ a_i \ |n_i) &= n_i^{1/2} |n_i - 1). \end{aligned} \quad (20)$$

Here one has to take into account all processes in thermal equilibrium corresponding to the same matrix element. $\bar{n}_i = (\exp(\hbar\omega_i/kT) - 1)^{-1}$ is the average thermal occupation number of phonons ω_i. In order to get net transition probabilities one has to subtract in Eq. (18) and (19) multiphononcreation and -annihilation processes. In the following, multiphonon processes and their connection with the temperature dependent halfwidth and the sideband structure of the localized mode main line will be discussed.

---

[2] $\Gamma(\omega)$ can not always be directly compared with $\Delta\nu$, the measured line width, since other effects such as strain broadening may come into play.

3.1.3. Temperature Dependence of the Main Line

The temperature dependence of the *halfwidth* of the main line is shown for the typical example of KCl with H_s^- and D_s^- centers in Fig. 8. For temperatures $< 50\,K$ the lifetime of the localized mode is determined by the decay into band modes. Because

$$\omega_L(D_s^-) < 2\omega_{max} < \omega_L(H_s^-) \tag{21}$$

(where ω_{max} is the highest band mode frequency; see Appendix), the D_s^- oscillator can decay into two band modes by $V_3(L,i,j)\ a_L a_i^+ a_j^+$ (diagram 3 in Table 6). For the temperature dependence of the two phonon decay process one obtains from Eq. (20)

$$(\bar{n}_i + 1)(\bar{n}_j + 1) - \bar{n}_i \bar{n}_j \approx 2\bar{n} + 1$$

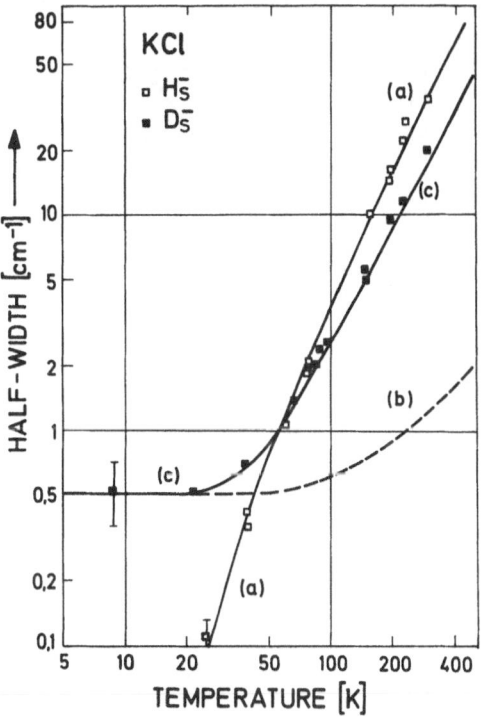

Fig. 8. Temperature dependence for the halfwidth of the main line. Curve a: Theoretical fit for H_s^- centers (phonon-phonon scattering according to Ref. [5]). Curve b: Temperature dependence of two phonon decay. Curve c: Theoretical fit for D_s^- line (phonon-phonon scattering plus two phonon decay) (after Ref. [51])

where $\bar{n}_i \approx \bar{n}_j = \bar{n}$ was put. The variation of $2\bar{n} + 1$ with temperature is shown in Fig. 8 curve (b). Because of Eq. (21) the H_s^- oscillator can decay at most in a four phonon process by a term $V_4(L,i,j,k)a_L a_i^+ a_j^+ a_k^+$ (see Eq. (17), and diagram 4 in Table 6). For the temperature dependence of this process one obtains in analogy $3\bar{n}^2 + 3\bar{n} + 1$. The absence of two phonon decay for the H_s^- oscillator explains the larger lifetime and thereby the smaller halfwidth of the H_s^- line with respect to the D_s^- line at low temperatures. Near 50 K a crossover of H_s^- and D_s^- curves occurs Above this temperature an elastic four phonon scattering process which was proposed by Elliott et al. [5] takes over (diagrams 2 and 5 in Table 6). This process gives a T^2 dependence above the Debye temperature and a T^7 dependence at very low temperatures. The best fit to the KCl: H_s^- data is obtained by using an effective Debye temperature of about 120 K [curve (a) in Fig. 8]. Furthermore, the scattering process is proportional to the square of the amplitude of the localized mode oscillator and thereby explains the factor two in the high temperature halfwidths for H_s^- and D_s^- lines. The theoretical curve (c) for the D_s^- line is obtained

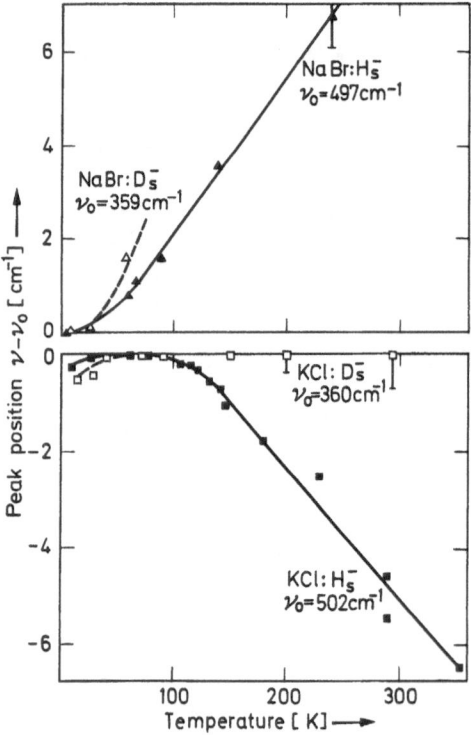

Fig. 9. Temperature dependent frequency shift of the main line (after Ref. [37])

by superimposing the two-phonon decay and the scattering contributions. This is done by adjusting the two phonon decay curve to the saturation halfwidth and adding curve (a) after multiplying it by a factor of 0.5.

The temperature dependent *frequency shift* of the main line was explained by Bilz et al. [51] as being due to the direct anharmonic damping and the thermal expansion of the lattice. The main contributions to the direct anharmonic damping are described by diagrams 1, 3, 6 and 7. Multiphonon processes of this type shift the harmonic localized mode frequency Ω_L to

$$\omega'_L = (\Omega_L^2 + 2\Omega_L \Delta_L(\omega))^{1/2}$$

where $\Delta_L(\omega)$ can be positive and negative as well. The thermal expansion of the lattice is caused by the anharmonicity of *all* band modes. It leads to a softening of the springs and thereby to a decrease in frequency. This additional contribution to the shift must be added to the above equation. It can be estimated from experimental results with mixed crystals [see Section 3.1.6, Eq. (26)] or from uniaxial stress experiments (see Section 3.1.7). The sum of contributions from direct anharmonic damping and thermal expansion may result in a positive or a negative frequency shift as observed for NaBr and KCl with U centers, respectively (Fig. 9).

3.1.4. Sidebands

The origin of the one phonon sideband spectrum is a three phonon process (see diagram 1 in Table 6). The localized oscillator is first *virtually* excited with light by a first order dipole moment. This intermediate state then decays into a final state in which a localized mode quantum is created and simultaneously a lattice phonon either created (sum process through terms $V_3(L, L, i) \, a_L a_L^+ a_i^+$) or annihilated (difference process through terms $V_3(L, L, i) a_L a_L^+ a_i$). The sideband peaks are located at an energy difference ω_i from the main line ω_L where

$$\omega_{\text{Light}} = \omega_L \pm \omega_i \, .$$

The difference process freezes out for $T \to 0$ because then $\bar{n}_i \to 0$. This can be seen from Fig. 5. The imaginary part of the susceptibility in the sideband region at $T = 0$ K may be written in the density approximation [51]

$$\chi''(\omega) = 16\pi^2 \left(\frac{3 M_1 V_3}{\omega_L \omega_i (1 + \omega_i/2\omega_L)} + M_2 \right)^2 \frac{\varrho(\omega_i)}{\omega_i \omega_L} \tag{22}$$

where $\varrho(\omega_i)$ is the one phonon density for the host crystal, and M_i are expansion coefficients for the dipole moment M. As mentioned in Section 3.1.2, M_2 can be neglected for the U center localized mode in

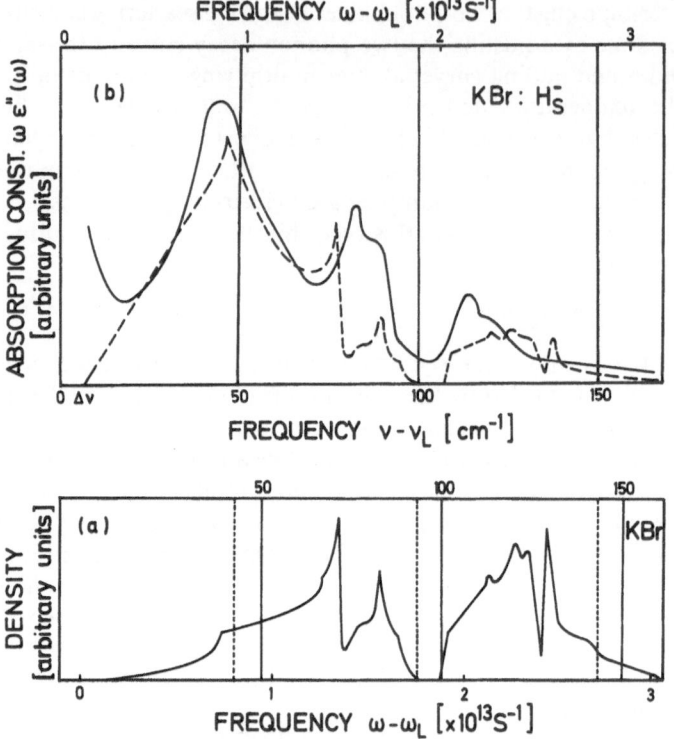

Fig. 10. (a) Phonon-density of states for KBr according to Cowley [83]. (b) Sideband absorption of the H_s^- center localized mode in KBr. Solid line: experimental curve by Fritz et al. [37]; Dashed curve: simple density approximation (see text)

alkali halides. A fit to the experimentally determined sideband spectrum using Eq. (22) is shown in Fig. 10. For $\varrho(\omega_i)$ the unperturbed one phonon density calculated by Cowley [83] was used.

Equation (22), however, is only an approximation, since out of all the possible phonon states, only those of special symmetry can couple to the localized oscillator. This can be verified by using the results of Section 2.2. There it was found that one has to consider only the initial state and the final state in order to find the selection rules. The initial state is that of the photons which transform like T_{1u} in O_h symmetry. The final state is described by $T_{1u} \times \Gamma$, where T_{1u} refers to the first excited state of the localized oscillator. In analogy to Eq. (5) one then finds that the only phonons which can couple to the localized oscillator are contained in the symmetric product

$$(T_{1u} \times T_{1u})_s = A_{1g} + E_g + T_{2g}. \qquad (23)$$

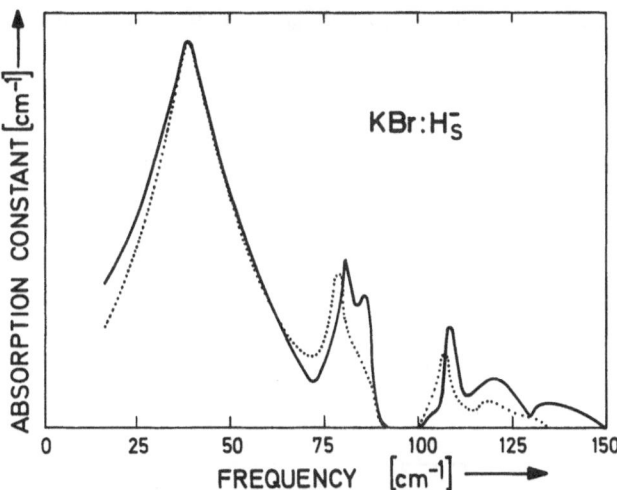

Fig. 11. High frequency sideband structure of the H_s^- center localized mode in KBr. Full line: theoretical curve (after Ref. [22]). Dotted line: modified experimental curve (after Ref. [60])

Their relative contribution to the sideband spectra depends on their coupling strength $A(\Gamma)$; i.e. instead of one "average" coupling parameter as in Eq. (22), three coupling parameters enter into the calculation.

By using these group theoretical aspects, Kühner and Wagner [22] calculated the sideband spectrum by employing the force constant model of Fig. 4. Their result is shown in Fig. 11 by the full line. It is a superposition of the imaginary parts of the perturbed projected Green functions $\hat{G}(\Gamma)$ (see Section 2.4) multiplied by $\omega^{-2}|A(\Gamma)|^2$. This line fits the modified experimentally determined curve (dotted line) much better than the simple density approximation according to Eq. (22). The symmetry vectors of 1 nn coordinates, which can couple to the localized mode are shown, among others, in Fig. 2. In the approximation of Ref. [22] the assumption is made that

$$A(A_{1g}) : A(E_g) : A(T_{2g}) = 1 : 1 : 1 . \tag{24}$$

Experimentally the anharmonic coupling coefficients can be determined from uniaxial stress experiments (see Section 3.1.5). For KBr : H_s^- it was found that the coupling of the localized mode to A_{1g}-type distortions is about a factor of three larger than the coupling to distortions of E_g and T_{2g} symmetry, i.e., the assumption of Eq. (24) is somewhat unrealistic. The local force constants were determined from the main line frequency and a best fit approximation of the sidebands. From this $A01 = -0.50$

Table 7. Local force constant changes $A01$, and $A14$ for H_s^- centers in alkali halides

Substance	$A01$	$A14$
NaF	-0.15	0
NaCl	-0.45	-0.18
NaBr	-0.44	-0.23
KCl	-0.43	0
KBr	-0.42	-0.2
KI	-0.48	-0.28

Force constant changes obtained from a shell model defect (see Fig. 4) calculation fitted to the H_s^- center localized mode frequency and to the sideband shape (after Ref. [46]).

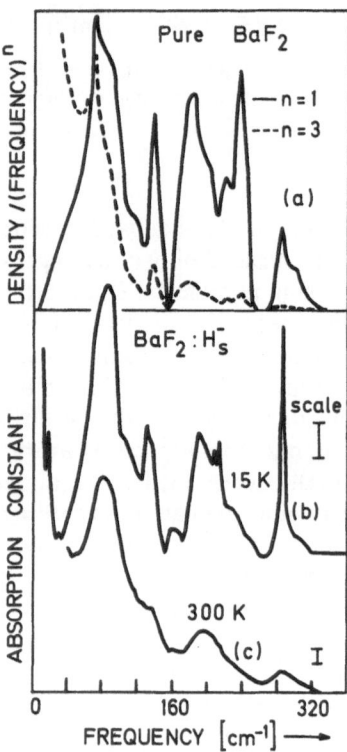

Fig. 12. (a) Phonon-density of states for BaF_2 as determined from neutron scattering at 300 K [84] weighted by ω_i^{-1} (solid line) and ω_i^{-3} (dashed line). (b) High frequency sideband structure of the localized mode fundamental in $BaF_2 : H_s^-$ ($6 \cdot 10^{18}$ cm^{-3}) at 15 K as observed by infrared absorption. (c) Same as (b) but at a temperature of 300 K. (after Ref. [66])

and $A14 = -0.13$ was obtained. Force constant changes $A01$ and $A14$ as obtained in a similar way from a shell model defect calculation by MacPherson and Timusk [46], are listed in Table 7. However, the curvature of the sideband structure does not react very sensitively to changes in $A14$. A much more accurate value of $A14$ is obtained by fitting the acoustic resonance frequency (see Chap. 4).

In spite of oversimplifications, like the assumption of Eq. (24) the remaining deviations between experimental and theoretical curves might reveal additional complications in the interpretation of the sideband structure. In this connection it was shown by Boese and Wagner [63] that the optical Jahn-Teller effect may affect the structural form of the sidebands, particularly in the low frequency region.

In alkaline earth fluorides the sideband structure of the hydrogen localized mode is not yet as well analyzed as for the case of alkali halides. The lattice modes which can couple to the localized mode, and which are observed in the sideband spectra, are of A_1, E, and T_2 symmetry. This is easily seen from Eq. (23) when employed for T_d symmetry. A characteristic sideband spectrum is shown in Fig. 12 (b) and (c) for $BaF_2 : H_s^-$. The sharp peak near $285 \, cm^{-1}$ is due to simultaneous excitation of the localized mode and a resonant mode associated with the H^- impurity. Fig. 12 (a) shows a simple density approximation. The curves represent the density of states according to Hurrell et al. [84] but are weighted by ω_i^{-1} (full curve), and ω_i^{-3} (dashed curve) respectively. In contrast to the localized mode sidebands in alkali halides it appears that a ω_i^{-1} weighting factor gives better agreement than a factor ω_i^{-3}!

3.1.5. External Fields

The dependence of the potential of the localized oscillator on changes in the positions of neighboring ions can be studied by stress experiments. Such experiments yield information on the anharmonic coupling coefficients, serve for determining the degree of degeneracy of a mode, and enable one, to study vibrational transitions which are forbidden by symmetry and which might become optically active under stress. Similar information can be derived from experiments with electric fields.

Uniaxial stress lifts the threefold degeneracy of the first excited level $(n = 1)$ of the localized oscillator. Fig. 13 shows the splitting of the main line in $KCl : H_s^-$ and $KI : H_s^-$ as a function of applied stress for different polarization directions. E is the vector of the incident light, P is the stress. The frequency shift of single line components is linear with stress. No changes in the band shape and the integrated absorption were found.

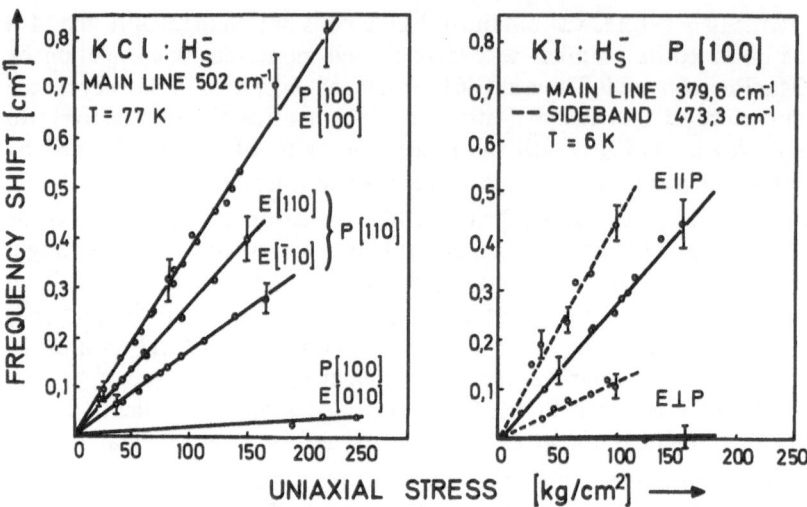

Fig. 13. Splitting of H_s^- center localized mode under uniaxial stress (after Ref. [43])

The strain Hamiltonian for a cubic crystal under uniaxial stress may be written in the form

$$
\begin{aligned}
H = P\{ &\alpha(A_{1g})\,(s_{11} + 2s_{12})\,r^2 \\
&+ \beta(E_g)\,(s_{11} - s_{12})\,[(2n^2 - l^2 - m^2)\,(2z^2 - x^2 - y^2) \\
&+ 3(l^2 - m^2)\,(x^2 - y^2)] \\
&+ \gamma(T_{2g})\,s_{44}(lmxy + lnxz + mnyz)\}
\end{aligned} \tag{25}
$$

for an arbitrary direction l,m,n of P relative to crystal axes. s_{ij} are *local* elastic stiffness constants [85]. The stress coefficients α, β, γ play the role of anharmonic coupling constants.[3] In O_h symmetry the strains associated with the coefficients α, β, γ transform like A_{1g}, E_g, and T_{2g}, respectively. In Eq. (25) only terms linear in the displacements of the neighbors and quadratic in the displacement of the H_s^- ion are included.

Stress in [110] direction lifts the degeneracy of the localized mode completely. The third component, which is polarized along [100] has not been measured for KCl: H_s^- (see Fig. 13). The strain energies as calculated from Eq. (25) are given in Table 8 for different polarization directions. From measured frequency splittings the stress coefficients α, β, γ can be calculated. They are listed for various alkali halides and for CaF$_2$ in Table 9. It is apparent from inspection of the stress coefficients

---

[3] Equation (25) was originally formulated by Gebhardt and Maier [28] in connection with the stress splitting of the electronic F band in alkali halides. There, the coefficients α, β, γ play the role of electron-phonon coupling constants in the optical transition.

Table 8. Energy shift per unit stress for different polarizations of stress P and electric vector of light E

Stress	Electric vector of light	Energy shift/stress
P	E	$-\Delta v/\Delta P$
hydrostatic	arbitrary	$3\alpha(s_{11}+2s_{12})$
001	001	$\alpha(s_{11}+2s_{12})+4\beta(s_{11}-s_{12})$
001	011	$\alpha(s_{11}+2s_{12})+\beta(s_{11}-s_{12})$
001	010	$\alpha(s_{11}+2s_{12})-2\beta(s_{11}-s_{12})$
110	110	$\alpha(s_{11}+2s_{12})+\beta(s_{11}-s_{12})+\frac{1}{2}\gamma s_{44}$
110	100	$\alpha(s_{11}+2s_{12})+\beta(s_{11}-s_{12})$
110	$1\bar{1}0$	$\alpha(s_{11}+2s_{12})+\beta(s_{11}-s_{12})-\frac{1}{2}\gamma s_{44}$
111	111	$\alpha(s_{11}+2s_{12})+\frac{2}{3}\gamma s_{44}$
111	$1\bar{1}0$	$\alpha(s_{11}+2s_{12})-\frac{2}{3}\gamma s_{44}$

Table 9. Anharmonic stress coefficients for high frequency localized modes due to substitutional hydrogen centers

Substance	$\alpha(A_{1g})$ $\times 10^4$ [erg/cm^2]	[cm^{-1}]	$\beta(E_g)$ $\times 10^4$ [erg/cm^2]	[cm^{-1}]	$\gamma(T_{2g})$ $\times 10^4$ [erg/cm^2]	[cm^{-1}]	References
KCl	$2{,}5\pm10\%$	840 (705)[b]	$0{,}8\pm10\%$	269 (165)	$0{,}35\pm10\%$	118 (118)	[43]
KBr	$2{,}0\pm10\%$	755 (565)	$0{,}64\pm10\%$	241 (151)	$0{,}58\pm10\%$	219 (219)	[72]
KI	$0{,}7\pm20\%$	308 (198)	$0{,}29\pm20\%$	128 (75)	$0{,}51\pm20\%$	224 (224)	[43]
CaF$_2$[a] (funda- mental)	$5{,}2\pm4\%$	905	$0{,}6\pm30\%$	104	$6{,}3\pm3\%$	1095	[82, 78]
KI (F center in-gap mode)		(125 ± 40)		(35 ± 15)			[86, 87]

[a] In alkaline earth fluorides where the U center has T_d site symmetry the stress coefficients transform like A_1, E, and T_2 respectively.
[b] Values in parantheses are corrected by Nardelli's formula.

that, in alkali halides, the coupling of the U center localized mode to A_{1g} distortions dominates. This shows the inadequacy of the assumption in Eq. (24). For CaF$_2$ the coupling to dilating A_1 and shearing T_2 distortions is comparable in size, and an order of magnitude larger than the coupling to distortions of E symmetry. Uniaxial stress experiments at the D_s^- line in CaF_2 were done by Hayes and MacDonald [78]. The corresponding stress coefficients are by about a factor $\sqrt{2}$ smaller than those for the H_s^- line. This again confirms that the H_s^- and D_s^- vibrations are highly localized. It will be seen in Section 3.2. that the U_1 center localized mode preferentially couples to T_2 lattice modes.

A correlation between the stress coefficients and the anharmonic coefficients of Eq. (13) is not straight forward, and a reliable derivation of such a relation is still missing. However, in alkali halides and alkaline earth fluorides, the approximate magnitude of the stress coefficients compares fairly well with estimates of the third order anharmonic coefficients (see Eq. (17)) as derived from the low-temperature linewidth of the D_s^- line or from the integrated sideband intensity [37, 5, 78].

The investigation of the behaviour of a sideband under stress was done for the 94 cm^{-1} in-gap mode in $KI:H_s^-$ (see Fig. 13) [43]. From the absence of a splitting of this band into more than two components, and from the fact that the relative frequency shift against the main line is practically independent of the polarization and the stress axis, one can conclude that this line is due to an A_{1g} vibration. This is in agreement with a calculation by Gethins et al. [23] which predicts an in-gap mode of A_{1g} symmetry as a consequence of the relaxed force constants pertinent to nearest neighbors of the U center.

In CaF_2, where H_s^- centers have T_d symmetry, stress induced transitions were observed [see Fig. 6(b)] [82]. This is due to the shear strain induced admixture of T_2 states with the infrared inactive levels of the excited $n = 2$ state. In alkali halides a similar activation of second harmonic levels should be possible under the influence of an external electric field.

The behaviour of the H_s^- center localized mode under an applied *electric field* was studied in CaF_2 [78, 82]. In this crystal, the electric field produces a strain of T_2 symmetry; and, as a consequence, a Stark splitting of the excited levels which is linear in the electric field strength was observed. It was found that fields of 10^5 V/cm, when applied in [111] direction, produce a splitting of at most 0.3 cm^{-1}.

3.1.6. Mixed Crystals

The localized mode absorption by U centers in alkali halide mixed crystals was investigated by Mirlin and Reshina [35] and by Barth and Fritz [42]. Fig. 14 shows the main features of the absorption for the example of $KCl:Rb$ with H_s^- centers. Besides the strongest line, ν_0, which has nearly the same position as in pure KCl with H_s^- centers, new lines $\nu_\alpha \ldots \nu_\varepsilon$ occur. On substituting H_s^- centers by D_s^- centers, the

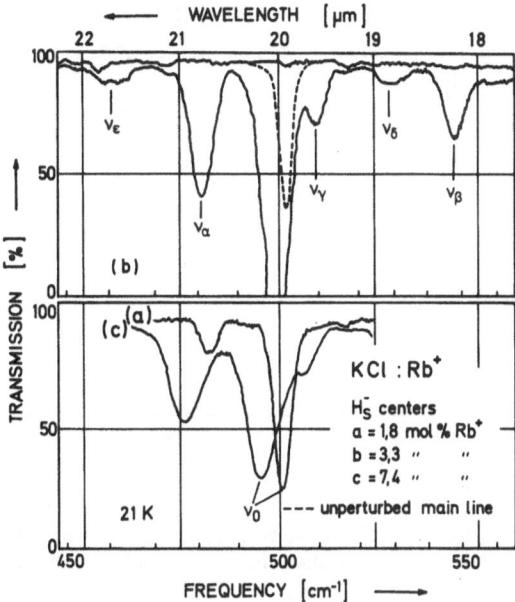

Fig. 14. Infrared absorption spectrum of H_s^- centers in KCl: Rb$^+$, at 21 K. v_0 is the nearly unperturbed H_s^- main line frequency, $v_\alpha \dots v_\varepsilon$ are new bands which are not seen in pure KCl (after Ref. [42]

additional lines undergo the same isotope shift as the main line, namely $v(H_s^-)/v(D_s^-) = 1.39 \approx \sqrt{2}$. In Ref. [42] lines v_α and v_β were interpreted as being due to U center configurations in which one of the 1 nn is replaced by the Rb$^+$ ion (see Fig. 32 in Section 4.2.2). The line v_γ is supposed to arise from a perturbation due to cations in the third shell. Lines v_δ and v_ε were correlated with U centers in the C_{2v} configuration [see case (d) in Fig. 32]. However, a further study of the relative intensities of lines seems to be necessary. They then could be compared, in analogy to Section (4.2.2), with relative frequencies in different configurations by taking into account the polarization and the degree of degeneracy in different modes.

Besides the frequency splitting the doping also causes a shift in the main line v_0. This is shown in Fig. 15. The decrease in frequency with increasing Rb$^+$ doping can be understood from an overall expansion of the lattice caused by the Rb$^+$ ions. Such an expansion of the lattice was verified from X-ray data by Gnaedinger [100]. The frequency shift can be compared with the results obtained from the uniaxial stress experiments reported in Section 3.1.5. From Table 8 one can easily derive the relation

$$\Delta v = (3 \Delta a/a) \, \alpha(A_{1g}).$$
(26)

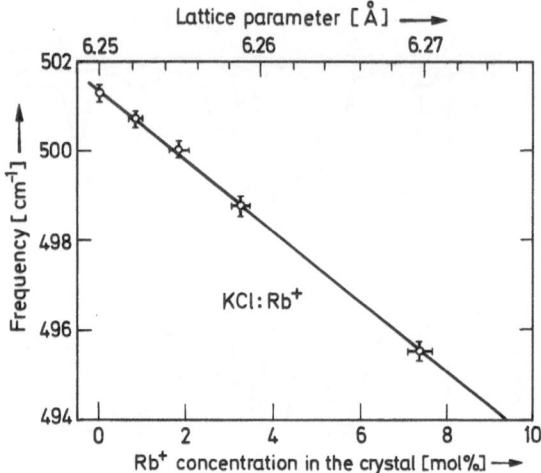

Fig. 15. Line positions v_0 for KCl:Rb$^+$ with H_s^- centers versus doping concentration at 21 K. The lattic parameter is interpolated using a Vegard relation known to hold at room temperature. The bands $v_\alpha \ldots v_\gamma$ show the same dependence (after Ref. [42])

From Fig. 15 one obtains $\alpha(A_{1g}) \approx 610 \, \mathrm{cm}^{-1}$. This value agrees fairly well with the value tabulated for KCl with H_s^- centers in Table 9, when one takes into account the fact that the lattice parameter change in the mixed crystal is about one order of magnitude larger than in the stress experiments.

3.1.7. Pair Centers

The near infrared absorption spectra of $H_s^- H_s^-$, $D_s^- D_s^-$, and $H_s^- D_s^-$ pair centers in KCl were studied by de Souza et al. [47]. For the aligned [110] pair centers in C_{2v} symmetry six non-degenerate localized modes are expected: three in phase and three out of phase modes which are polarized in [110], [1$\bar{1}$0] and in [001] directions. The out of phase modes are infrared inactive for $H_s^- H_s^-$ and $D_s^- D_s^-$ pairs, but active for the mixed $H_s^- D_s^-$ centers. The infrared active in phase modes will be polarized in the longitudinal (L), and in the transverse (T_1 and T_2) directions of the pair. This is shown in the insert of Fig. 16. Using polarized light, the infrared measurements of the localized mode spectrum of crystals containing [110] aligned pairs revealed the position and polarization of the pair modes. The spectra are shown for the case of $H_s^- H_s^-$ pair centers in Fig. 16. Three lines at 463.5, 512.5, and 535 cm^{-1} are found

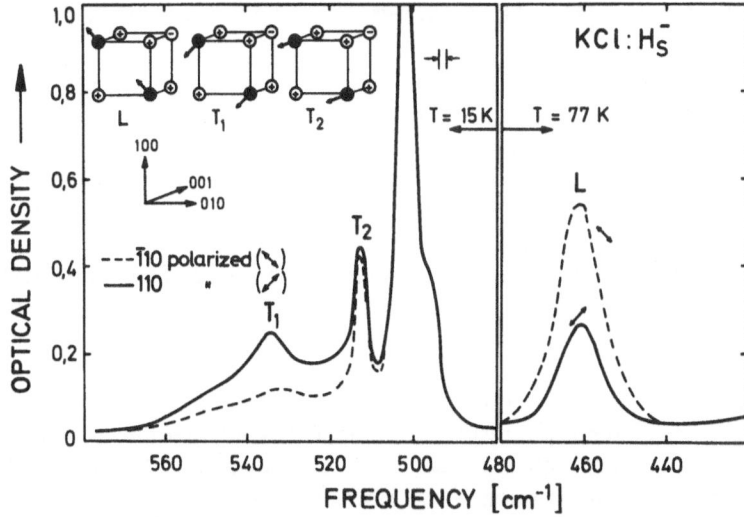

Fig. 16. Absorption spectrum of KCl: H_s^- ($3 \cdot 10^{18}$ cm^{-3}) after production of aligned H_s^- H_s^- pairs (with preferential [$\bar{1}$10] orientation as indicated in the upper left-hand corner), measured for [$\bar{1}$10] and [110] polarized light incident in a [001] direction. The phonon sideband of the H_s^- localized mode at 565 cm^{-1} has been subtracted. The band at 463 cm^{-1} was measured at 77 K in order to avoid errors from insufficient resolution (after Ref. [47])

Table 10. Measured and calculated line positions for pair centers (after Ref. [88])

		$H_s^- H_s^-$		$D_s^- D_s^-$		$H_s^- D_s^-$	
		exp.	calc.	exp.	calc.	exp.	calc.
[110]	in phase	463	463	331	329	351	347
	out of phase		537		380	508	507
[1$\bar{1}$0]	in phase	535	535	375	379		348
	out of phase		465		330	511	506
[001]	in phase	512	512	368	363		353
	out of phase		492		349		503

which can be correlated with L, T_2 and T_1 modes, respectively. The measured line positions for $H_s^- H_s^-$, $D_s^- D_s^-$ and $H_s^- D_s^-$ centers are listed in Table 10 together with calculated values which are based on a linear chain model.

3.2. U_1 Centers (H_i^-, D_i^- Centers)

3.2.1. Perturbed U_1 Centers

In analogy to U centers, U_1 centers give rise to localized mode absorption in the near infrared. However, in alkali halides, the interaction with nearby anion vacancies (α centers) – which are simultaneously produced in the $U - \alpha$ process (see Section 2.1) – results in a point symmetry of U_1 centers which is in general lower than cubic (T_d).

The near infrared vibrational spectrum, which was first investigated by Fritz [69], essentially consists of three groups of lines (see Fig. 17). Single groups of lines can be correlated with H_i^- centers of different thermal stability. The interaction between U_1 centers and α centers is of elastic and electric (Coulomb) nature. Its origin is the charge of the defects. This interaction results in the tendency for thermal recombination of U_1 centers with α centers to U centers (see Fig. 18), and in the splitting of the absorption lines. Both, the thermal instability of U_1 centers, and the amount of splitting in the vibrational spectrum, are inversely proportional to the relative distance between the defects.

It is not completely understood why essentially three distinct pairs of vacancy – interstitial ion configurations exist. From the observation of three line groups and three relative sharp annealing steps one can conclude that the interstitial ions occupy in each stage a single site or

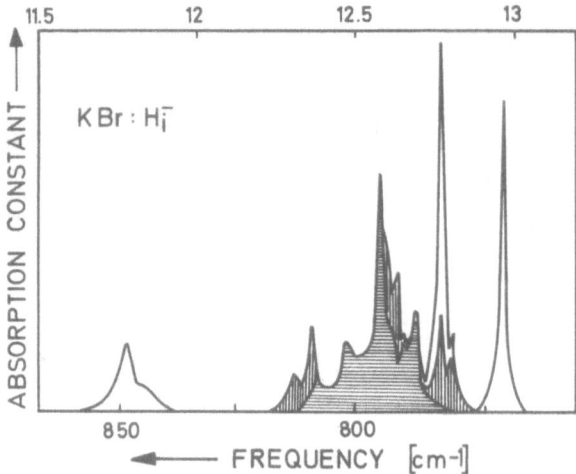

Fig. 17. Infrared absorption spectrum due to H_i^- ions in KBr after low-temperature (20 K) irradiation in the ultraviolet U band. The various line groups disappear after warming to the following temperature regions: white, $90-110$ K; vertical striping, $130-160$ K; horizontal striping, $190-230$ K (after Ref. [69])

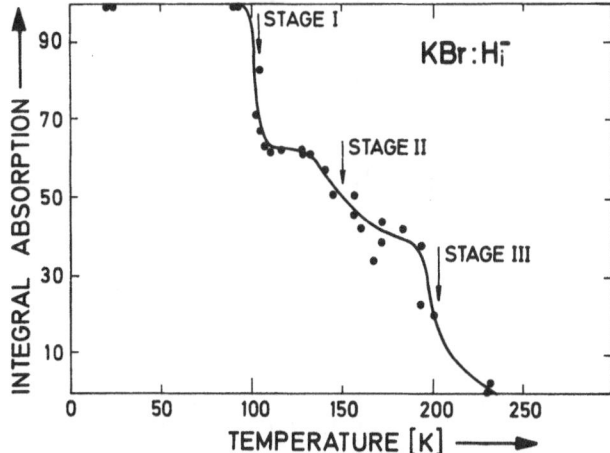

Fig. 18. Decrease of the total integral absorption in the spectrum of Fig. 17 as a function of annealing temperature (after Ref. [69])

a small number of almost equivalent sites. Notice that the U_1 centers undergoing annealing between 190 and 230 K (stage III) still feel the influence of the vacancy. However, it was shown recently for the case of KBr and KI that a careful thermal annealing procedure leads to H_i^-, and D_i^- centers, whose point symmetry is *not* lowered by short-range interaction with nearby anion vacancies. The symmetry of these "free" H_i^-, D_i^- centers turns out to be cubic (T_d).

3.2.2. The Main Line

The near infrared vibrational spectra of nearly "free" H_i^- and D_i^- centers in KBr and KI are shown in Figs. 19 and 20. In analogy to the near infrared vibrational absorption of U centers, the spectrum consists of a main line and sidebands. Table 11 summarizes some of the experimental data. Preliminary results on the main line frequencies in LiF, KCl, RbCl, and RbBr are included.

The main line shows a temperature dependent half width which follows a T^2 law above about 80 K. When cooling the sample to liquid helium temperature in KBr : H_i^- and KI : H_i^- the line width becomes $\lesssim 1.5 \, \text{cm}^{-1}$ (see Table 11). This remaining half width might still be influenced by a "small" interaction with vacancies. However, no splitting in the main line was observed! This leads to the conclusion that for free U_1 centers the hydrogen moves in a cubic potential of T_d symmetry; i.e.

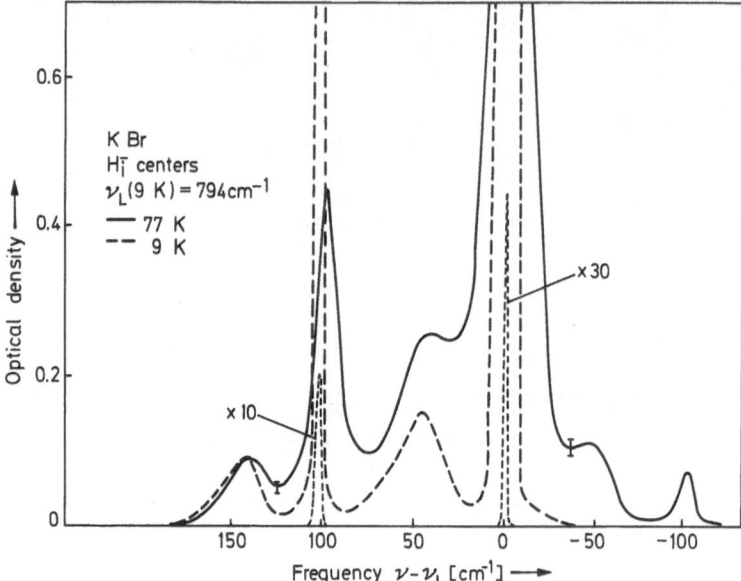

Fig. 19. Analyzed near-infrared spectra of "free" H_i^- centers in KBr at 77 and 9 K (after Ref. [72])

the H^- or D^- ion is sitting in the middle of a double tetrahedron formed by anions and cations, respectively. The vibrational transition then occurs from a non-degenerate A_1 ground state to a three-fold degenerate $T_2(1)$ excited state. A model for this vibration is included in Fig. 41. It is in contrast to the model suggested by Gross and Bron [70]. The strong localization of the U_1 center mode is seen in the isotope shift which is approximately $\sqrt{2}$ for H_i^- and D_i^- centers as expected from the mass ratio (see Table 11 and Sections 2.3 and 3.1.1). An absorption due to second or third harmonic transitions, which are allowed by symmetry (see 3.1.1), was not yet found. From the measurements one can conclude that the intensity of these transitions is $\lesssim 10^{-3}$ when compared with the A_1 to $T_2(1)$ transition.

Interstitial hydrogen ions were found also in CaF_2, but only when the crystals were doped with rare earth ions. Fig. 21 shows the infrared absorption spectra of rapidly quenched hydrogenated CaF_2 crystals containing Gd ions. The spectra show four lines. Two of these lines, one at $965\,cm^{-1}$ – which was observed also in pure CaF_2 (see Section 3.1) – and one at $1296\,cm^{-1}$, have frequencies which are independent of the rare earth ions. From a number of experiments (see Ref. [7]) and the fact that no higher harmonics could be observed, the

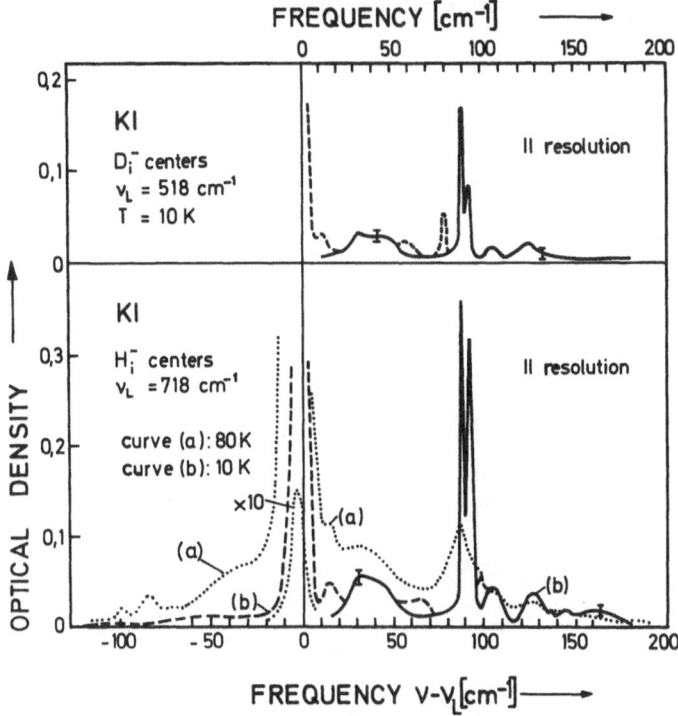

Fig. 20. Near infrared spectra of H_i^- and D_i^- centers after "purification" by annealing. Broken lines: experimental curves. Full lines: analyzed sideband shapes (after Ref. [71])

Table 11. Main data on the high frequency localized modes and in-gap modes due to interstitial hydrogen centers in alkali halides and alkaline earth fluorides

Substance	Localized mode frequency $\nu_L(H_i^-)$ [cm^{-1}]	Halfwidth $\Delta\nu_L(H_i^-)$ [cm^{-1}]	Freq. ratio $\nu_L(H_i^-)/\nu_L(D_i^-)$	Temperature [K]	Intensity ratio I_{SB}/I_T	In-gap mode frequency ν_G [cm^{-1}]	References
LiF	2100			130			[73, 74]
KCl	850			20			[69]
KBr	794	<1.5	1.40	9	0.20	98.7	[69, 72]
KI	718	<0.8	1.39	10	0.15	86.7	[71, 86]
RbCl	800			20			
RbBr	748			55			
CaF$_2$	1310	15		77			
	1296	30	1.39	300			[15]

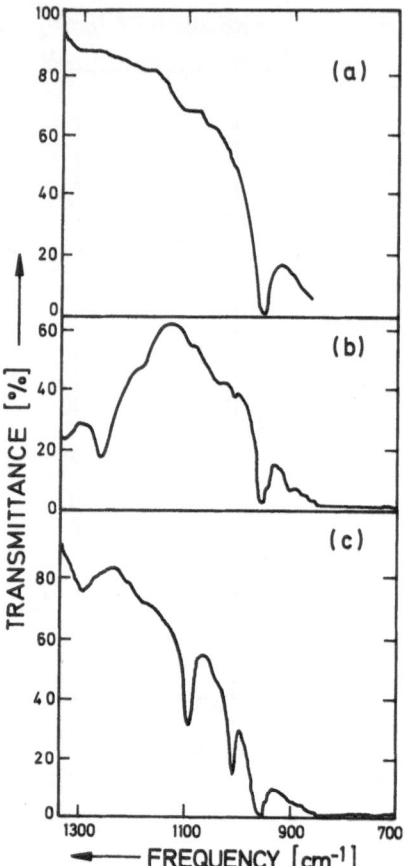

Fig. 21. Infrared absorption spectra of (a): hydrogenated pure CaF_2. (b): slowly annealed (20°/h) hydrogenated CaF_2 containing Gd. (c): rapidly quenched hydrogenated CaF_2 containing Gd (after Ref. [7])

$1296 \, \text{cm}^{-1}$ line was correlated with hydrogen ions which are well separated from rare earth ions and which are sitting on an interstitial position at the center of an empty fluorine cell. These hydrogen ions have O_h site symmetry. A model is shown in Fig. 22. In deuterated crystals a similar line occurs near $932 \, \text{cm}^{-1}$. The ratio of absorption frequencies is 1.39 (see Table 11). The frequencies of the remaining two lines at $1007 \, \text{cm}^{-1}$ and $1093 \, \text{cm}^{-1}$ in the spectrum of Fig. 21 vary with the particular rare earth ion present. These lines could be correlated with interstitial hydrogen ions adjacent to rare earth ions which split the first excited T_2 level of the hydrogen oscillator twofold. The site

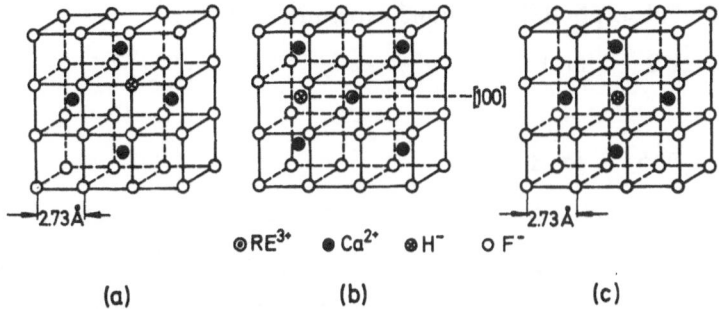

Fig. 22. Structure of U_1 centers in CaF_2. (a): pure CaF_2. (b): the axial-site center. (c): the cubic-site center (after Ref. [7])

symmetry of these hydrogen ions is C_{4v}. It was established from the number of second harmonics (three) observed, and from ESR measurements on crystals containing Ce and Nd ions respectively.

3.2.3. Sidebands

For an identification of the one phonon sidebands, U_1 center concentrations of some $10^{17}/cm^3$ are necessary. With these "high" concentrations it was not possible to completely suppress a small amount of perturbed U_1 centers. However, one can identify the one phonon sideband spectra by using the following criteria:

a) The frequency displacements of sideband peaks should be equal for H_i^- and D_i^- centers.

b) The peaks should be roughly symmetric about the main line; the high and low frequency peaks have temperature dependent intensities according to $\bar{n}+1$ and $\bar{n}$, respectively (see Sections 3.1.2 and 3.1.4).

Figure 19 shows the sideband spectrum for $KBr:H_i^-$ which was identified according to these criteria (see also Ref. [72]). The results for $KI:H_i^-, D_i^-$ are shown in Fig. 20. The broken lines indicate a "rest" of perturbed U_1 centers. The frequencies of the absorption lines arising from perturbed H_i^- and D_i^- centers are in a ratio $\nu_L(H_i^-)/\nu_L(D_i^-) \approx \sqrt{2}$. Such a ratio is expected for perturbed main lines (see Section 3.1.6).

The phonon sidebands arise through anharmonic coupling between the localized mode and lattice band modes – in analogy to the U center localized mode sidebands. The low frequency sidebands disappear for $T \to 0$ (see 3.1.4). For the ratio of integral absorption of sidebands and the total absorption it was found experimentally

$$I_{SB}/I_T \approx 0.2 \quad \text{for KBr}$$

and   (27)

$$I_{SB}/I_T \approx 0.15 \quad \text{for KI.}$$

This ratio is a measure of the amount of anharmonic coupling [37, 51]. It is roughly the same as for the high frequency localized mode of U centers [37]. This means that the total third order anharmonic coupling is about the same for U and U_1 centers. However, the U_1 center localized mode couples preferentially to one mode, say Γ, whose frequency in the sideband spectrum occurs in the band gap. This can be seen from Figs. 19 and 20. The ratio of integral absorption of this single line and the total sideband absorption is experimentally

$$I_\Gamma/I_{SB} \approx 0.5 \quad \text{for KBr}$$

and

$$I_\Gamma/I_{SB} \approx 0.2 \quad \text{for KI} \tag{28}$$

where I_Γ for KI contains only the low energy component of the doublet.

From Eq. (23), when employed to T_d symmetry, one can easily verify that the only lattice modes which can couple to the localized mode, are of A_1, E, or T_2 symmetry. Among the T_2 modes are polar modes T_2^P as well as rhombic distortions T_2^S (shear modes). The fact that polar T_2^P modes are infrared active would suggest the possibility of a search for these modes by means of far infrared spectroscopy; the results of such investigations are reported in Section 4.6.

3.3. U_2 Centers (H_i^0, D_i^0 Centers)

The only results on interstitial hydrogen atoms in alkali halides were published for LiF [73]. In γ and neutron irradiated LiF : OH⁻ a line at 2200 cm⁻¹ is observed which was correlated with the localized vibration of H_i^0 centers. It would be nice, however, if one could further check the results, e.g. by starting out from LiF : OD⁻ material.

Low temperature X-ray or UV irradiation of hydrogenated CaF_2 crystals containing rare-earth ions converts hydrogen ions on interstitial lattice positions into interstitial neutral hydrogen atoms. This was established for centers having C_{4v} symmetry (see Section 3.2.2) by ESR and ENDOR measurements in crystals containing Ce and Nd ions respectively [7]. The existence of a localized mode due to interstitial hydrogen atoms appears from the vibronic spectra. Vibronic transitions separated from their parent lines by $767\,\text{cm}^{-1}$ and $565\,\text{cm}^{-1}$ were observed for hydrogenated and deuterated crystals, respectively. Direct infrared localized mode absorption of interstitial hydrogen atoms in CaF_2 occurs near $640\,\text{cm}^{-1}$ [75]. The effective charge associated with this vibration is about 0.07 electrons (compared to 0.7 to 0.9 electrons for U centers in alkali halides).

3.4. Interstitial Hydrogen Molecules ($H_{2,i}^0$ Centers)

Akhvlediani and Politov [73] observed in γ and neutron irradiated $LiF:OH^-$ crystals a line at $2000\,\text{cm}^{-1}$ which they believe as being connected with interstitial hydrogen molecules. A further clarification of the experimental situation seems to be necessary (see Section 3.3).

4. Infrared Vibrational Absorption: In-Gap Modes and Resonant Modes

In-gap modes and acoustic resonant modes of electron and hydrogen centers were studied experimentally only in the alkali halides KBr and KI. Table 12 summarizes the infrared vibrational frequencies. From Table 12 and Table A1 in the Appendix, one finds that $U(H_s^-, D_s^-)$ and F' centers give rise to acoustic resonant mode absorption ([18, 72, 86, 89–91]; [91]), $F, F_A(\text{Na})$, and $U_1(H_i^-, D_i^-)$ centers to in-gap mode absorption ([72, 86, 87, 91]; [86]; [71, 72, 86]). The only theoretical investigations were performed for U centers [89, 92], and for F centers [87, 94, 95].

In analogy to high frequency localized modes, the frequency positions and splitting of in-gap modes and acoustic resonant modes allow conclusions on local force constants and on local site symmetries of a defect.

For the case of U centers and U_1 centers, where high frequency localized mode absorption *and* acoustic resonant mode or in-gap mode absorption is observed, one can try to calculate the far infrared line frequencies and their absorption shape by using only the parameters determined from the near infrared spectra. The comparison of calculated and measured far infrared spectra yields information on the adequacy and the sensitivity of a lattice dynamical model.

For the case of U_1 centers the lack of inversion symmetry allows the direct far infrared excitation of phonons, which can be excited indirectly in the localized mode sideband spectra as well. From the comparison of near and far infrared spectra one can experimentally determine the

Table 12. In-gap mode and resonant mode absorption of electron and hydrogen centers in alkali halides

Substance	Frequency $\nu_{G,R}$ [cm^{-1}]	Half-width $\Delta\nu_{G,R}$ [cm^{-1}]	Temperature [K]	References
KBr				
$:H_s^-$	89 ± 0.5		7	[89, 90]
$:D_s^-$				
$:F$	99.60 ± 0.03			
	99.07 ± 0.04	<0.1	1.2	[87, 90]
	98.50 ± 0.05			
$:H_i^-$	98.7 ± 0.5	<1.8	9	[72]
$:D_i^-$	98.7 ± 0.5	<1.8	9	[72]
KI				
$:H_s^-$	61 ± 0.5			[18, 89, 91]
$:D_s^-$	60.50 ± 0.5			[89]
	60.25 ± 0.5			[71]
$:F$	82.62 ± 0.02			
	81.98 ± 0.02	<0.1	1.2	[91, 87]
	81.19 ± 0.05			
$:F_A$(Na)	(82)	<1.2	4.2	[86]
	80.05			
$:M$	—			[86]
$:F'$	68.5 ± 0.5	<1	7	[91]
$:\alpha$	—			[86]
$:H_i^-$	86.7 ± 0.5	<1.3	7	[71]
$:D_i^-$	86.7 ± 0.5	<1.3	7	[71]

relative magnitude of anharmonic coupling coefficients for phonons of special symmetries.

For F centers extensive experimental and theoretical studies of the electronic absorption band, which shows strong temperature-dependent broadening due to electron phonon coupling, did allow the prediction of some lattice dynamical properties [1, 93] Vice versa, the direct investigation of the vibrational spectra yields a better understanding of the electronic transition, and of the electron phonon coupling. The fact that the electronic absorption band does not exhibit distinct vibrational structure, but is smooth and nearly Gaussian (strong coupling case), allows one to obtain a reasonable quantitative picture of the band shape by employing oversimplified dynamical models for the lattice – as e.g. the configurational diagrams (see e.g. Ref. [96]). However, only a detailed treatment of the impurity-lattice vibrational problem can reproduce the shape of the electronic F band quantitatively without making use of fitted ad hoc parameters. Such a calculation may, in addition, supply insight into the electron-phonon energy exchanges.

This treatment allows one to determine from the infrared absorption and Raman spectra, local force constants which may serve for a calculation of local elastic relaxations. These relaxations, in turn, affect the overlap between the wave functions of the electron and its neighbors, and must therefore be taken into account in a calculation of the electronic energy of the defect.

Both, electron and hydrogen centers may give rise to defect induced absorption, which yields information on host lattice band mode frequencies and band mode densities.

4.1. U Centers (H_s^-, D_s^- Centers)

Difference spectra of the far infrared absorption of KBr and KI containing U centers are shown in Figs. 23 and 24. The absorption bands are located in the acoustic phonon band of the host lattice. Even at low temperatures their linewidth remains large. The substitution of H_s^- centers by D_s^- centers in KI results only in a shift of the central component of the threefold structured line near 61 cm^{-1}. This central line shifts to

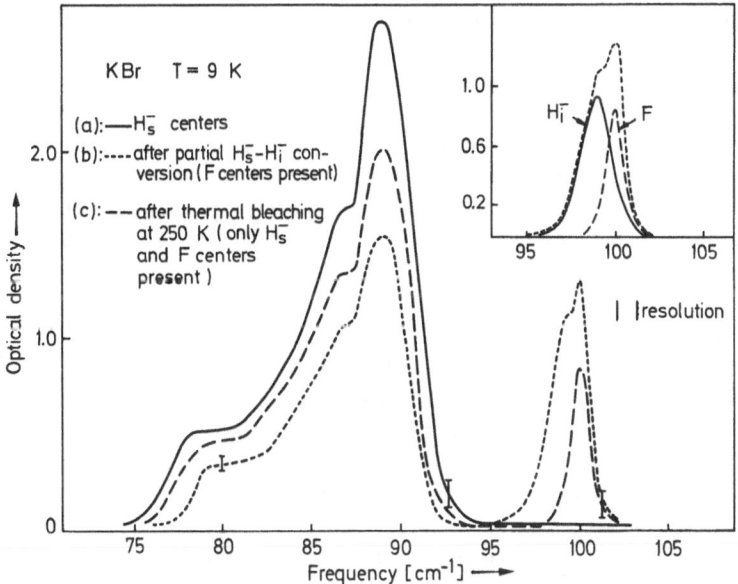

Fig. 23. Acoustic resonant mode absorption of H_s^- centers in KBr (full line). Broken line shows the result of the $U - \alpha$ process: H_s^- centers are converted to H_i^- centers and α centers (a small amount of F centers is simultaneously produced). H_i^- centers and F centers lead to in-gap mode absorption (after Ref. [72])

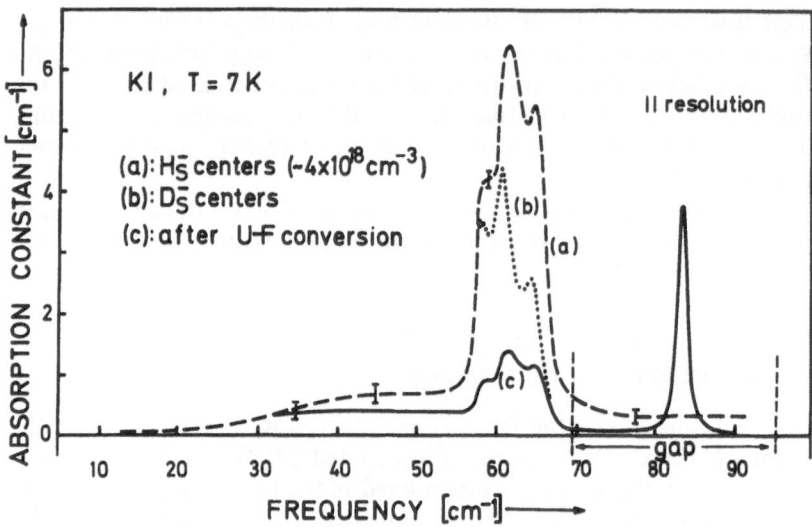

Fig. 24. Acoustic resonant mode absorption of H_s^- centers (broken line) and D_s^- centers (dotted line) in KI. Full line shows the result of the $U-F$ conversion (after Ref. [71])

longer wavelengths by about 0.75 cm^{-1}. Only this line can be correlated with a $T_{1u}(2)$ vibrational mode of the hydrogen ion (see Figs. 2 and 3). The remaining structure is the defect induced *one – phonon* band mode absorption.

The absorption of originally inactive host lattice band modes yields information on their density of states; e.g. one finds that the maxima and discontinuities occuring in the absorption spectra outside of the resonance do not depend upon the type of defect. As an example, the U center induced acoustic band mode absorption in KBr is compared with the absorption induced by Na$^+$ and Cl$^-$ centers (see Fig. 25). The frequency positions of discontinuities (labelled by arrows) coincide for vanishing defect center concentrations. They can be correlated *directly* to singularities in the phonon spectrum of the host lattice (see last row in Table 13 and Fig. 10a). The character of the singularity, however, may change. This can be concluded from a comparison with phonon density of states curves as derived from neutron inelastic scattering experiments (see Fig. 10a).

In the $T_{1u}(2)$ resonant mode, the hydrogen moves nearly in phase with 1 nn (see Figs. 2 and 3). The vibrational amplitude is *not* localized at the defect site, in fact, the amplitudes at 1 nn sites and at the defect site are of about the same magnitude. One can approximately describe this vibration by a rigid oscillator – consisting of the defect and 1 nn –

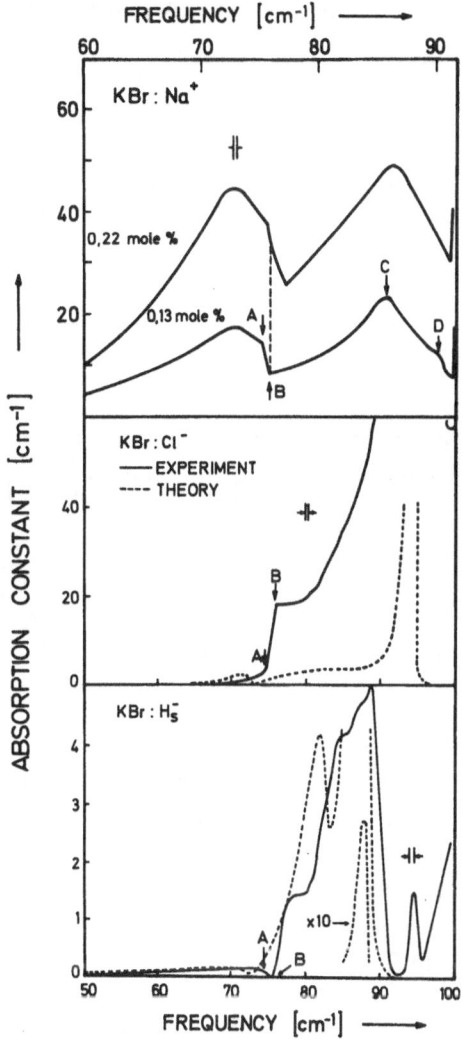

Fig. 25. Acoustic band mode absorption induced by Na^+, Cl^-, and H_s^- centers in KBr (see also Fig. 23). Discontinuities in the spectra are marked by A, B, C, and D. Note that in the case of Cl^- the slope has reversed (after Ref. [98])

vibrating against 4 nn (see Section 2.3). Therefore, the isotope shift expected is small. In KI : H_s^-, D_s^- the ratio of the resonance frequencies found experimentally is

$$v_R(H_s^-)/v_R(D_s^-) \approx 1.01$$

Table 13. Positions of observed singularities in KBr crystals (see Fig. 25) with H_s^-, Cl^-, and Na$^+$ centers at 11 K

Defect	Mol% defect	A [cm^{-1}]	B [cm^{-1}]	C [cm^{-1}]	D [cm^{-1}]
H_s^-	0.001	75	75.5	85	86
Cl$^-$	2	74.7	76	–	–
Na$^+$	0[a]	74.68	75.2	85.22	89.73
Shell model		70.7	71.3	83	88

[a] Extrapolated to zero concentration.

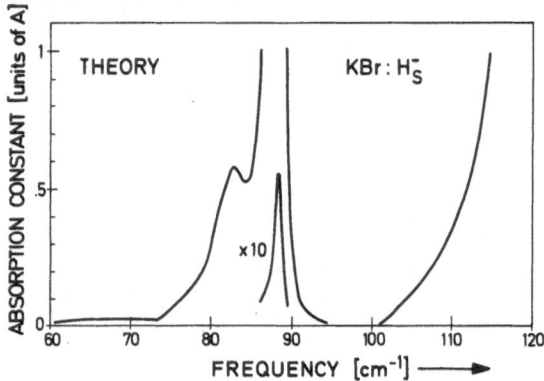

Fig. 26. Calculated absorption spectrum for KBr : H_s^- [The unit A is $4\pi e^2\, p(n^2(\infty) + 2)^2/ 9n(\omega)\,\mu c \times 10^{-12}$ sec] (after Ref. [89])

A three dimensional model calculation on the far infrared absorption of KBr and KI with U centers was performed by Woll et al. [89]. These authors employ the same force constant model, which they earlier used for a theoretical fit of the high frequency localized mode absorption (see Section 3.1). This model enables one to predict the far infrared absorption spectra without introducing any further parameters. The results are shown in Figs. 26 and 27. The comparison with the experimental curves in Figs. 23 and 24 shows that the overall shape of the spectrum is reproduced fairly well. The very sharp peak in both of the theoretical curves is the resonance. The fact that these peaks are narrower than the experimental ones might be due to the neglect of anharmonic contributions. Furthermore, discrepancies occur in the exact resonance positions, especially for KI. These discrepancies could be due either to faults in the neutron-derived shell-model density of states or to inadequacies in the

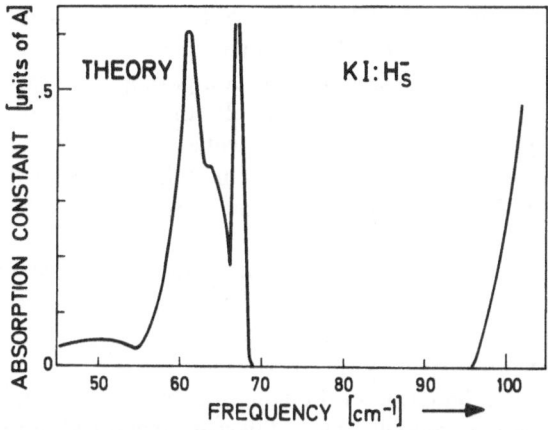

Fig. 27. Calculated absorption spectrum for $KI : H_s^-$ (see caption of Fig. 26 for unit A.) Model evidently predicts the resonance too high by $6-10\ cm^{-1}$ (after Ref. [89])

force constant model for the defect. From the results in Section 4.2, it is believed that the latter is mainly responsible for the discrepancies. The far infrared mode frequencies are extremely sensitive to force constant changes A14 – much more than the sideband shape. A shell model fit of the force constant changes A01 and A14 to the main line frequency of the localized mode and the resonant mode frequency, respectively, yields

$$A01 = -0.43; \quad A14 = -0.20 \text{ for KBr}$$

and

$$A01 = -0.48; \quad A14 = -0.23 \text{ for KI}.$$

In spite of the reservations which should be made when one compares local force constants (see Section 4.2.4), the deviation in A14 from the value which was obtained by Kühner and Wagner by fitting the shape of the localized mode sideband for $KBr : H_s^- (A14 = -0.13)$ is obvious.

4.2. F Centers

4.2.1. In-Gap Mode and Acoustic Band Mode Absorption

According to the simple model of Section 2.3 one would expect for F centers a vibrational mode of the type $T_{1u}(2)$. The frequency of this mode should lie in the acoustic optical band gap. In this vibration the electron is "carried along" by 1 nn. In fact, for F centers this vibration was found experimentally. The results of the $U \rightarrow F$ conversion in KBr and KI are included in Figs. 23 and 24 respectively.

These spectra were obtained by comparing the transmission of "doped" and "undoped" samples. In KBr the F center mode is found at the upper gap-edge near 100 cm^{-1}, while a similar line in KI occurs almost in the middle of the gap near 83 cm^{-1}. In both cases the linewidth is determined by the resolution of the spectrometer which is about 1 cm^{-1} in these experiments.

A detailed theoretical investigation of the lattice dynamics of the F center in connection with the band shape of the electronic absorption, the F center induced first-order Raman spectra (see Chap. 5), and the infrared absorption spectra was performed by Benedek and Mulazzi (1969) [94]. The calculation of these authors is based on one of Hardy's deformation dipole models. Two different dynamical force constant models are suggested. The infrared absorption spectra calculated from

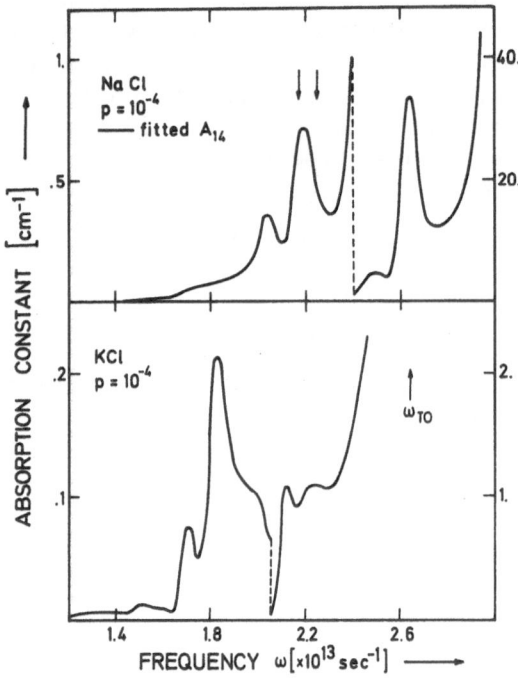

Fig. 28. Infrared absorption spectra due to F centers in NaCl and KCl as obtained for the extended dynamical model. The arrows indicate the frequencies for which the resonant denominator is zero. Note that the vertical scale is changed in the highfrequency region near the reststrahl frequency ω_{TO} (after Ref. [94])

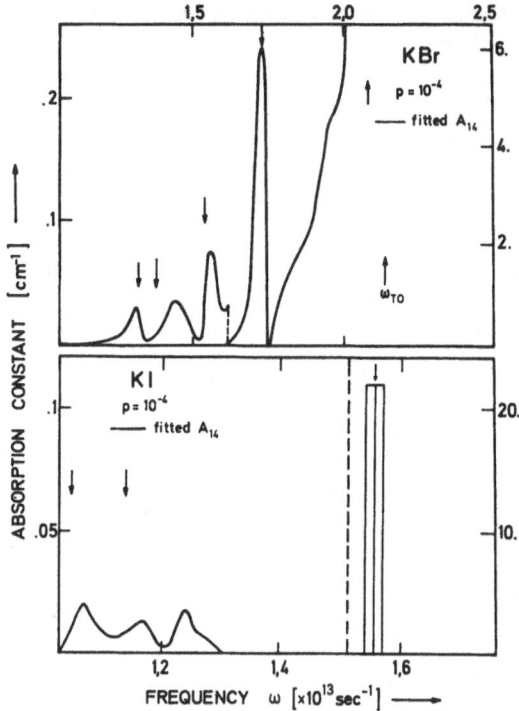

Fig. 29. Same as Fig. 28, but for KBr and KI. In both spectra the induced T_{1u} in-gap modes and their strengths are evident (after Ref. [94])

one of their models, the one which corresponds to the model of Fig. 4, are shown in Figs. 28 and 29. The force constants A_{01} and A_{14} are fitted to the experimental values for the halfwidth of the electronic F band and the Huang – Rhys parameter for zero temperature. The concentration of F centers was assumed to be $p = 10^{-4}$ (see Section 2.4). All calculated spectra show defect induced band mode absorption. Zeros in the resonance denominator (see Section 2.4) are indicated by arrows. In NaCl they appear in the region between the acoustical and transverse optical modes, where the host lattice phonon density turns out to be large. In this case the absorption shape therefore essentially exhibits the T_{1u} projected density rather than resonant mode structure. The band mode absorption due to F centers was not yet verified experimentally.

The spectra for KBr and KI show strong in-gap modes (at $90 \, \text{cm}^{-1}$ and $82 \, \text{cm}^{-1}$ respectively) whose frequencies are in agreement with the experimentally determined frequencies within 10%. The remaining deviations in the exact line positions are mainly due to the fact that the force constant fit mentioned above is not very sensitive to changes in A_{14}. A very sensitive method for obtaining local force constants A_{01} and A_{14} from the infrared data, independent from data on the electronic transition, is reported in Section 4.2.2.

A calculation on the F center induced in-gap mode absorption in NaBr, NaI, KBr and KI was performed by Singh and Mitra [95]. These authors use a model, earlier suggested by Jaswal [97], in which the ions are assumed as being rigid, and where the impurity and its six 1 nn are treated as a vibrating molecule.

The force constant – in analogy to A_{01} – was fitted to the frequency of the electronic F band, thereby employing the picture that the electronic absorption band is in reality the local mode of the F center electron (see beginning of Chap. 3). The calculated frequency position of the $T_{1u}(2)$ in-gap mode for KBr and KI is $92.6 \, \text{cm}^{-1}$ and $81.2 \, \text{cm}^{-1}$, respectively. In-gap modes in NaBr and NaI should occur near $120 \, \text{cm}^{-1}$ and $103 \, \text{cm}^{-1}$, respectively.

4.2.2. The Isotope Splitting

Figs. 30 and 31 show absorption spectra of the F center in-gap mode for KBr and KI with a resolution of about $0.1 \, \text{cm}^{-1}$. In both cases a threefold structure with lines A, B, and C is revealed (with respect to the line near $80 \, \text{cm}^{-1}$ in KI see Section 4.3). These lines are extremely sharp. Their real half width is not yet resolved. The relative intensities of these lines are constant and independent of the method of F center production. No frequency shift in lines A and B was observed on changing the F center concentration between 10^{17} and $3.10^{18} \, \text{cm}^{-3}$.

The ratio of the integral absorption, $\int \alpha(v) \, dv$, of lines A, B, and C is experimentally for the case of KI

$$I(A):I(B):I(C) = 1000 : (145 \pm 10\%) : (5 \pm 25\%) . \tag{29}$$

For KBr one obtains

$$I(A):I(B):I(C) = 1000 : (162 \pm 20\%) : (8 \pm 50\%) . \tag{30}$$

This is within the error of the analysis, the same as for the case of KI.

In a series of experiments it was verified that the threefold structure of the F center in-gap mode is not due to the presence of impurities or due to F aggregate centers (see Ref. [87]). This leads to the conclusion that this structure is due to the presence of two stable potassium isotopes

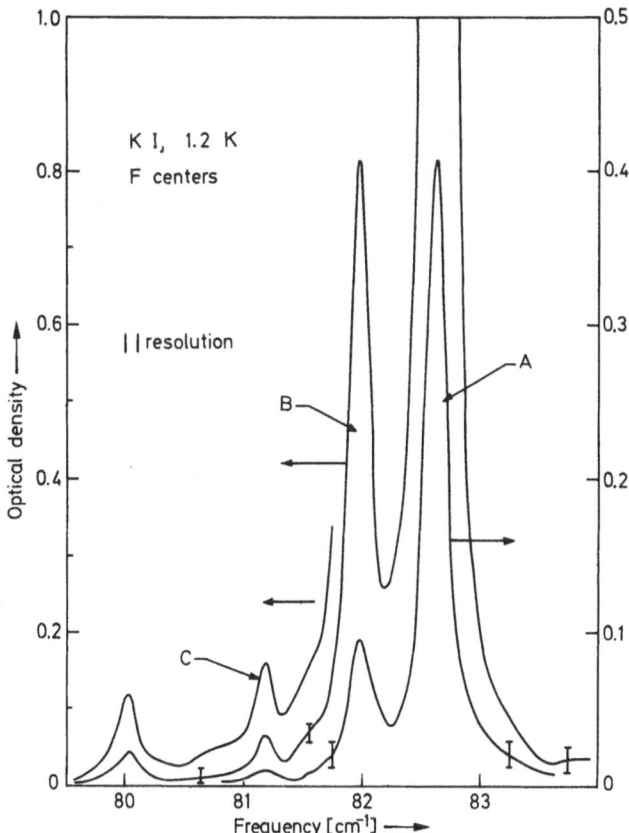

Fig. 30. F center in-gap mode in KI measured with a resolution of 0.1 cm^{-1} (apodized spectrum). The positions of the lines are: $A = 82.62 \pm 0.02$ cm^{-1}; $B = 81.98 \pm 0.02$ cm^{-1}; $C = 81.19 \pm 0.05$ cm^{-1} (after Ref. [87])

in the crystal, K^{39} and K^{41}. The natural abundance ratio of these two isotopes is K^{39} : K^{41} = 93 : 7.

One now can make the following model: The F center has three different configurations of symmetry O_h, C_{4v}, and D_{4h} (see Fig. 32). In the first case, (a), the F center is surrounded only by potassium isotopes K^{39}. Therefore, one obtains a vibrational transition from a non-degenerate A_{1g} ground state to a threefold degenerate T_{1u} excited state. The latter splits into a twofold degenerate E and a non-degenerate A_1 state when one replaces one of the 1 nn by the heavier isotope, K^{41}, thereby reducing the point symmetry to C_{4v} [case (b)]. Substituting two of the 1 nn lying on opposite sites of the F center one ends up with D_{4h} symmetry [case (c)]. The degeneracy is the same as for case (b). For case (c) the

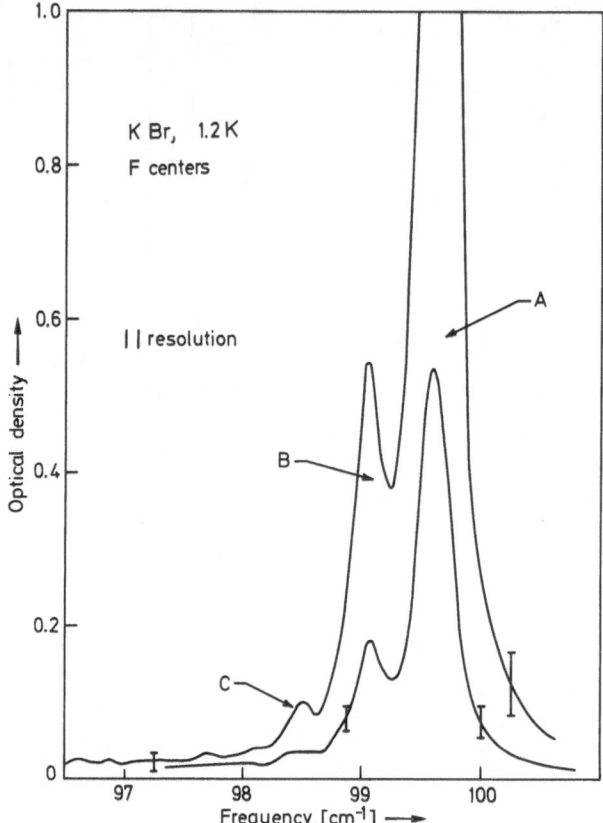

Fig. 31. Same as Fig. 30, but for KBr. The positions of lines are $A = 99.60 \pm 0.03 \text{ cm}^{-1}$; $B = 99.07 \pm 0.04 \text{ cm}^{-1}$; $C = 98.50 \pm 0.05 \text{ cm}^{-1}$ (after Ref. [87])

splitting of the excited levels, however, is larger because of the bigger perturbation in effective masses.

One now can assume that cases (a) to (c) differ only in the mass perturbation; i.e. it is assumed that the local force constants are the same for all three cases. Then one can reduce *all* possible configurations in the crystal to cases (a) to (c). As an example, case (d) then does not lead to additional vibrational transitions. The B_1 and E, and the B_2 and A_1 vibrations respectively are degenerate. From this model one expects three vibrational transitions A, B, and C. One now can calculate their relative intensities: The total number of F centers shall be k. Denoting the probability that $n F$ centers have $i \, K^{41}$ ions as 1 nn by $P_i(n)$, the average occupation number is then

$$b(i) = \Sigma_n P_i(n) = k\binom{m}{i}(1 - 1/k)^{m-i} k^{-i}. \tag{31}$$

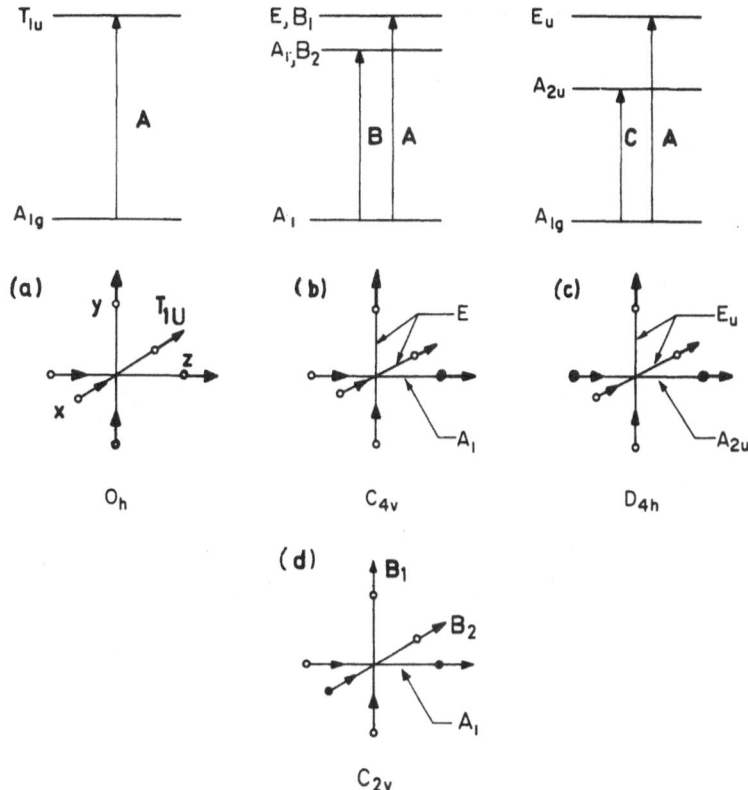

Fig. 32. Vibrational transitions for different F center configurations. Case (a): K^{39} environment only. Case (b): one of the 1 nn substituted by K^{41}. Case (c): two 1 nn on opposite lattice sites substituted by K^{41}. Case (d): two 1 nn on *non*-opposite lattice sites substituted by K^{41}

For the natural abundance ratio of $K^{39}:K^{41}=93:7$. One finds among $6k$ potassium ions $m=6k\cdot7\cdot10^{-1}$ of the heavier isotope. One then obtains

$$b(0)=65.5\%;\quad b(1)=27.7\%;\quad b(2)=5.9\%.$$

In order to compare the relative frequencies of these three configurations with the relative intensities of the lines, one must take into account the polarization and the degree of degeneracy in the modes according to Fig. 32. One then obtains

$$A=2b(0)+(4/3)\,b(1)+(12/15)\,b(2),$$

$$B=(2/3)\,b(1)+(16/15)\,b(2),$$

$$C=(12/15)\,b(2).$$

The above values then yield

$$I(A) : I(B) : I(C) = 1000 : 143 : 5.$$

This is in excellent agreement with the experimental data (29) and (30). This results supports the model and encourages a threedimensional model calculation.

4.2.3. Model Calculation

The calculation is based on the model of Fig. 4. Changes in the tangential force constant B are neglected. For $s = 0, 1, 4$ the symmetry vectors (see Section 2.4) are $\sigma(T_{1u}, j, 0)$; $\sigma(T_{1u}, j, 1, 1)$; $\sigma(T_{1u}, j, 4, 1)$ where $j = 1, 2, 3$.

In the following the projected perturbation matrix $\hat{V}(T_{1u})$ for configurations (a) to (c) of the F center according to Fig. 32 is discussed.

Case (a): F center in O_h symmetry

$$M' < M_A ; \qquad \varepsilon_- = 1 - M'/M_A$$

$$A01 \neq 0 ; \qquad A14 \neq 0.$$

The perturbation matrix $\hat{V}$ can be written

$$\hat{V} = \begin{bmatrix} \hat{V}_x & & \\ & \hat{V}_y & \\ & & \hat{V}_z \end{bmatrix}$$

V_α are symmetrical 3×3 matrices of the form

$$\hat{V}_\alpha^{(a)} = \begin{bmatrix} \varepsilon_- \cdot \omega^2 + A \cdot A01 \cdot M_A^{-1} & -\sqrt{2}\,A \cdot A01 \cdot (M_A M_K)^{-1/2} & 0 \\ \dots\dots\dots\dots\dots\dots & A(A01 + A14) \cdot M_K^{-1} & A \cdot A14 \cdot (M_K M_A)^{-1/2} \\ \dots\dots\dots\dots\dots\dots & \dots\dots\dots\dots\dots\dots & A \cdot A14 \cdot M_A^{-1} \end{bmatrix}$$

and

$$\hat{V}_x^{(a)} = \hat{V}_y^{(a)} = \hat{V}_z^{(a)}.$$

In this case all gap modes are threefold degenerate.

Using the assumption of Section 4.2.2 ($A01, A14$ the same for the three configurations) the perturbation matrix can be written directly down for cases (b) and (c).

Case (b): F center in C_{4v} symmetry

$M' > M_K;$ $\varepsilon_+ = 1 - M'/M_K$

where M' and M_K are masses of K^{41} and K^{39} isotopes respectively

$\hat{V}_x^{(b)} = \hat{V}_y^{(b)} = \hat{V}_x^{(a)}$

$\hat{V}_z^{(b)} = \hat{V}_z^{(a)} + \begin{bmatrix} 0 & & \\ & \frac{1}{2}\varepsilon_+ \cdot \omega^2 & \\ & & 0 \end{bmatrix}.$

One gets two types of modes: the unperturbed modes of the first model (twofold degenerate) and one non-degenerate new mode.

Case (c): F center in D_{4h} symmetry

$\hat{V}_x^{(c)} = V_y^{(c)} = \hat{V}_x^{(b)} = \hat{V}_x^{(a)}$

$\hat{V}_x^{(c)} = 2\hat{V}_z^{(b)} - \hat{V}_z^{(a)} = \hat{V}_z^{(b)} + \begin{bmatrix} 0 & & \\ & \varepsilon_+ \cdot \omega^2 & \\ & & 0 \end{bmatrix}$

The degeneracy is the same as for case (b), but the splitting is larger. No further gap modes can occur in connection with the heavier mass isotope K^{41}. All T_{1u} modes which strain the springs between 1 nn and 2 nn correspond to the case of a diatomic linear chain with $M_K < M_A$ and $\varepsilon_+ < 0$, where no gap modes exist. For a more complete discussion see Refs. [87] and [20].

4.2.4. Numerical Results for Frequency Positions

Figures 33 and 34 show the calculated frequency dependence of lines A, B, and C of the F center in-gap mode in KI and KBr as a function of force constant changes $A01$ and $A14$. The results are in agreement with the predictions in Section 2.3: the line positions depend strongly on $A14$, the force constant change between 1 nn and 4 nn. For any *fixed* value of $A01$ the peaks may be shifted over the total gap region by a variation of $A14$ within reasonable limits ($-0.3 \le A14 \le 0.3$). On the other hand the "rigid oscillator model" in which the F center and 1 nn vibrate against 4 nn, is too simple. The frequency positions depend also on $A01$ – but by a much smaller amount.

The threefold structure of the F center in-gap mode allows a quantitative determination of the force constant pairs, $A01$ and $A14$, which is

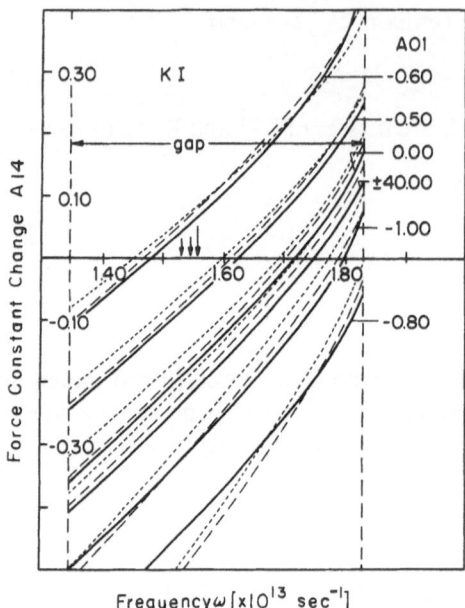

Fig. 33. Dependence of the line position and splitting of the F center in-gap mode absorption in KI as a function of force constant changes $A01$ and $A14$. Solid line: F centers in the $b(0)$ configuration. Dashed line: F centers in the $b(1)$ configuration. Dotted line: F centers in the b (2) configuration. The calculation was done for 0 K. The arrows show the experimental positions of lines (after Ref. [87])

independent of data from the electronic F band: a fit of the observed frequency for the line A in KBr and KI restrict the possible force constant pairs $A01$, and $A14$. A further restriction is obtained by fitting the measured splitting of lines A and B. The frequency of line C is best reproduced by choosing values as listed in Table 14. The positive sign for $A14$ for KBr is at the limit of accuracy of the calculation. However, one can see the tendency of $A14$ to decrease when going from KBr to KI. In contrast, $A01$ remains roughly constant!

A comparison of local force constant changes for F centers (Table 14) and U centers (Table 7) is not straight forward. The reason is that the force constant changes may "largely" depend on the model used for calculating host lattice eigenvalues and eigenvectors. Keeping this in mind one can at most speak about the tendency of local force constant changes when substituting U centers with F centers: From Tables 7 and 14 one finds that values $A01$ for U centers and F centers may be comparable (this was also found in the calculations by Singh and Mitra [95]; see also Section 4.2.1) and only "slightly" different for different substances

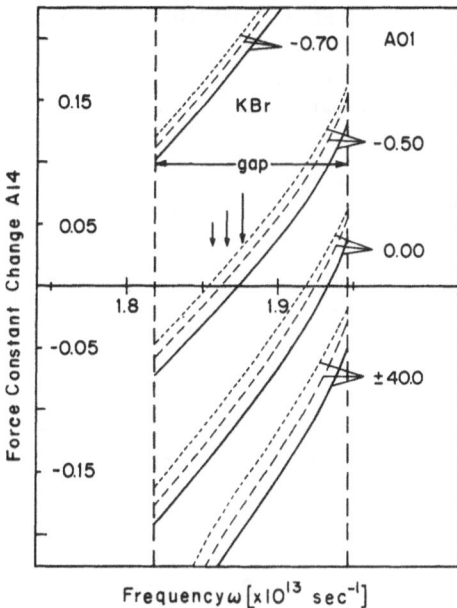

Fig. 34. Same as Fig. 33, but for KBr (after Ref. [87])

Table 14. Local force constant changes $A01$ and $A14$ for F centers in alkali halides

Substance	$A01$	$A14$
KBr	− 0.50	+ 0.002
KI	− 0.50	− 0.060

Force constant changes obtained from a defect model calculation − based on Schröders breathing shell model-fitted to frequencies of lines A, B, and C of the F center in-gap mode.

(the case of NaF is excluded). On the contrary, values $A14$ seem to be much smaller for U centers than for F centers, and decrease with increasing size of halide ions. A possible heuristic explanation may be the following: The force constant $A14$ is determined by the relaxation of 1 nn, while $A01$ is mainly due to the "vacancy", i.e. an additional contribution due to relaxation is small. The relaxation of 1 nn ions is governed by short range Born-Mayer repulsive forces and by Coulomb forces. The repulsive forces are smaller between the defect and 1 nn than that between 1 nn and 4 nn. This leads to an inward relaxation of 1 nn ions. The Coulomb forces are determined by the charge within the first shell. For U centers the charge is completely localized within the first shell.

For *F* centers, however, there is a depletion of charge due to the locally large extended ground state, i.e. a positive "defect charge" within the first shell occurs. This results in a Coulomb repulsion, i.e. an outward relaxation of 1 nn, and thereby in an increase of $A14$. This would explain the smaller values $A14$ for *U* centers relative to the values for *F* centers. In addition the larger value of $A14$ for *F* centers in KBr as compared to that for KI could be explained in this manner. The depletion of charge is larger with KBr than with KI [the lattice constants are a (KBr) = 3.28 Å, a(KI) = 3.48Å]. For KCl with *F* centers one would expect the value of $A14$ to be even larger than for KBr. This is supported by an analysis of ENDOR measurements by Kersten [99], which yields the relaxation of 1 nn. From the simple model suggested in Ref. [87], one calculates with Kersten's data $A14 = +0.072$ for KCl.

A further support of these arguments is obtained from experiments with charged defects like *F'* centers (see Section 4.5).

4.2.5. Stress Experiments

The stress splitting of lines *A* and *B* of the *F* center in-gap mode in KI is shown in Fig. 35 for $E \| P \| [100]$ and $E \| [001]$, $P \| [100]$. *E* is the electric vector of the incident light, and *P* the stress. Each line splits into two components which shift linearly with applied stress but by different amounts. Within the accuracy of measurements no change in the band shape or the integral absorption ($\pm 30\%$) was observed. From the slope of the function $\Delta v_{1,2}(A) = f(P)$, one can, for different directions of polarization, determine the stress coefficients α and β (see Section 3.1.5).

Uniaxial stress in the [100] direction lowers the point symmetry for *F* centers in the $b(0)$ configuration from O_h to D_{4h}. In the $b(1)$ configuration the symmetry becomes C_{4v} for the *c* axis $\| P$ and C_{2v} for the *c* axis $\perp P$ (see Fig. 36). Centers in the $b(2)$ configuration shall be neglected in the following. However, because no measurable splitting was observed between the A_{2u} and B_1 levels and among the E_u and *E* and B_2 levels, one can assume that the static distortions are nearly of A_{1g}, E_g, and T_{2g} symmetry.

The local compliances were calculated according to Benedek and Nardelli [85]. With the force constant changes $A01$ and $A14$ determined in 4.2.4 one obtains

$$s_{11} + 2s_{12} = 3.05 \times 10^{-12} \, \text{cm}^2/\text{dyn} \,,$$

$$s_{11} - s_{12} = 5.00 \times 10^{-12} \, \text{cm}^2/\text{dyn} \,.$$

From Eq. (25) one calculates for line *A*:

$$\alpha(A_{1g}) = 125 \pm 40 \, cm^{-1}; \quad \beta(E_g) = 35 \pm 15 \, cm^{-1} \,. \tag{32}$$

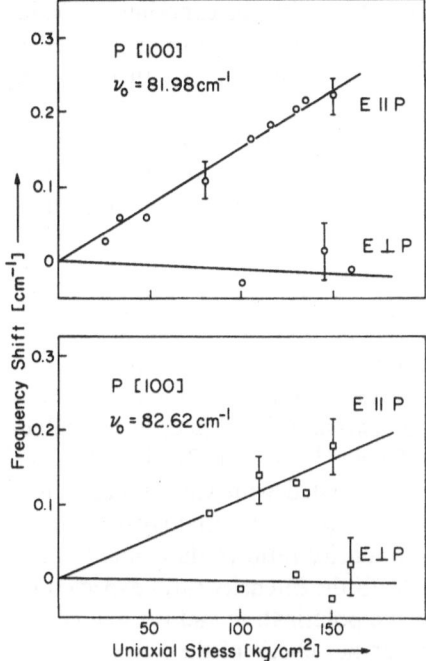

Fig. 35. Frequency shift $\Delta v = v - v_0$ of the lines $A(v_0 = 82.62 \text{ cm}^{-1})$ and $B(v_0 = 81.98 \text{ cm}^{-1})$ of the F center in-gap mode in KI (5 K) with uniaxial stress (after Ref. [87])

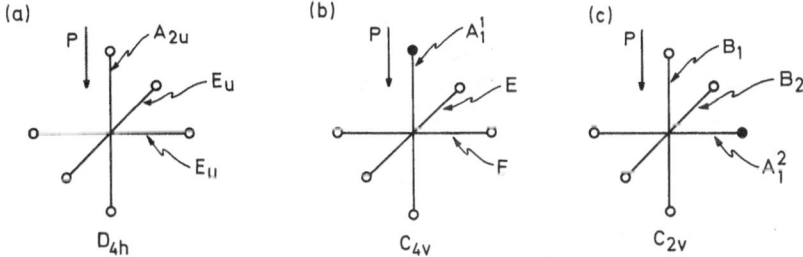

Fig. 36. Behaviour of F centers in the $b(0)$ [case (a)] and $b(1)$ [cases (b) and (c)] configurations with uniaxial stress. ○ denotes K^{39} and ● denotes K^{41} isotopes. The single components of the split lines A and B contain

$P \parallel E : v_1(A) : A_{2u} ; \ B_1$

$\qquad v_1(B) : A_1^1$ "shifted".

$P \perp E : v_2(A) : E_u ; E ; B_2$

$\qquad v_2(B) : A_1^2$ "slightly shifted" (after Ref. [87])

This result is consistent with the experimental values for line B. The excited vibrational state of the F center is coupled about three times more strongly to spherical distortions A_{1g} then to tetragonal distortions E_g. This ratio is similar to that for U centers.

Because the half widths of lines A, B, and C of the F center in-gap mode were not yet resolved, an estimate of the third order anharmonic coefficient – which could be compared with the stress coefficients – is not possible.

4.3. F_A(Na) Centers

Preliminary experiments have been performed on the vibrational absorption of F_A(Na) centers [86] (see Section 2.1). Besides the unperturbed F center in-gap mode two additional lines in the difference spectra were found at $80.05 \pm 0.04 \text{ cm}^{-1}$ and at $72.61 \pm 0.04 \text{ cm}^{-1}$ [4]. In principle these lines could be correlated with vibrational transitions $A - A_1$ and $A_1 - E$ [see Fig. 32 case (b) – in analogy with electronic transitions F_{A1} and F_{A2}]. The intensity ratio of these additional lines seems to be in rough agreement with the intensity ratio expected from Eq. (31) when taking into account the polarization and the degree of degeneracy in the modes. However, a further check of the correlation of these lines with F_A(Na) centers is necessary. Most illuminating would be experiments on the dichroism at the F_A center far infrared lines.

4.4. M Centers

In additively coloured crystals F centers were converted to M centers (at 300 K by irradiation of the crystal with F center light). No sharp absorption line of similar oscillator strength, when compared with the F center mode, was observed in the gap.

4.5. Infrared Vibrational Absorption of F' and α Centers

In the preceeding sections defects whose valency is the same as that of the ion replaced have been discussed.

In this chapter the influence of the effective charge within the "vacancy" on the infrared vibrational frequency $T_{1u}(2)$ will be studied – at least experimentally. This can be done by converting F centers to F' centers and α centers, or U centers to U_1 centers and α centers (see Fig. 1, and Section 2.1).

---

[4] In Na doped KI, in addition to the defect induced acoustic band mode absorption earlier seen by Sievers [18], an in-gap mode was found at 83.81 cm^{-1}, which is due to a localized mode of Na^+ centers.

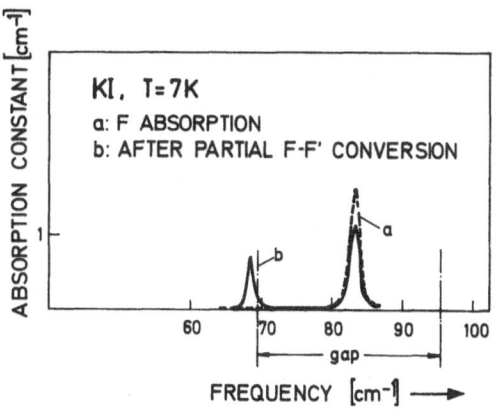

Fig. 37. $F - F'$ conversion in KI at 7 K (after Ref. [91])

The result of the $F-F'$ reaction in KI is shown in Fig. 37. A fraction of about 30% of F centers was converted to F' centers and anion vacancies, α centers. In the infrared spectra a new resonant mode at the upper edge of the acoustic band near 68.5 cm^{-1} occurs. This line is due to F' centers[5] This could be verified by using the $U-\alpha$ process (see Section 2.1). This reaction leads to a new absorption line near 87 cm^{-1}, which can be correlated with interstitial hydrogen ions, U_1 centers (see Section 4.6). No absorption occurs at 68.5 cm^{-1}. α centers do not absorb in the frequency region investigated.

Qualitatively, one can understand these spectra from the results obtained in Section 4.2. An additional charge within the vacancy attracts 1 nn while 4 nn are repelled [see Fig. 38 (a)]. This results in a decrease in the force constant $A14$. According to Fig. 33 one therefore expects a lower frequency for the vibrational absorption of F' centers when compared with the frequency of the F center in-gap mode.

The anion vacancy behaves as a defect charge. Therefore, one has the opposite conditions [see Fig. 38 (b)]. 1 nn are repelled while 4 nn are attracted. This results in an increase in $A14$. An absorption due to α centers is therefore expected at a higher frequency relative to the F center in-gap mode. Because no absorption due to α centers was found one can assume that this absorption is covered by the intrinsic absorption of the host lattice.

For a quantitative analysis of the line position of the infrared vibrational absorption due to F' centers, one can no longer assume trans-

---

[5] The instrumental resolution in this experiment was about 1 cm^{-1}. A high resolution measurement on this line — in analogy to the measurements on the F center in-gap mode (see Section 4.2.2) — would be of great interest for a quantitative treatment of the problem.

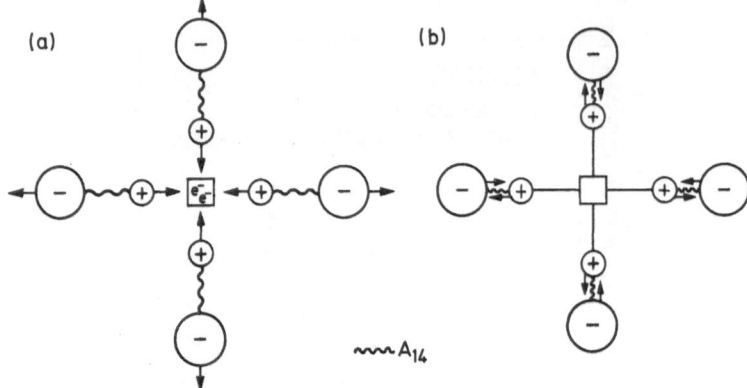

Fig. 38. Model for relaxation of 1 nn of the F' center [case (a)] and the α center [case (b)]. Arrows indicate directions of relaxation

lational symmetry in the matrix of effective charges. Furthermore, it seems to be doubtful whether a dynamical model which is limited to a small region around the defect (see Section 2.4) is appropriate at all. On the contrary, one would assume that the additional charge within the vacancy causes Coulomb potential changes which extend far into the lattice. Under these aspects a theoretical treatment of the F' center vibrational problem seems to be exciting.

4.6. U_1 Centers (H_i^-, D_i^- Centers)

In the far infrared spectra of KBr and KI the $U - \alpha$ process yields additional absorption lines at $98.7 \pm 0.5\,\mathrm{cm}^{-1}$ and at $86.7 \pm 0.5\,\mathrm{cm}^{-1}$, respectively (see Figs. 23 and 39). Because F centers are simultaneously produced (see Section 2.1), the F center in-gap mode is also present in both spectra. The correlation of the additional absorption lines with U_1 centers was established by using the $F-F'$ reaction (see 4.5). Furthermore, the lines bleach along with the ultraviolet U_1 band, which was controlled in the same experiment. The thermal stability of these centers corresponds to "free" U_1 centers (see Section 3.2). The half widths of the lines are $\lesssim 1.8 \pm 0.5\,\mathrm{cm}^{-1}$ and $\lesssim 1.3 \pm 0.5\,\mathrm{cm}^{-1}$ for KBr and KI, respectively. Substituting H_i^- centers for D_i^- centers the same lines were obtained. Within the accuracy of the measurement no frequency shift was found.

The nature of the U_1 center in-gap mode is quite different from the F center in-gap mode or the U center resonant mode. No simple explanation on the basis of an altered mass and altered force constants

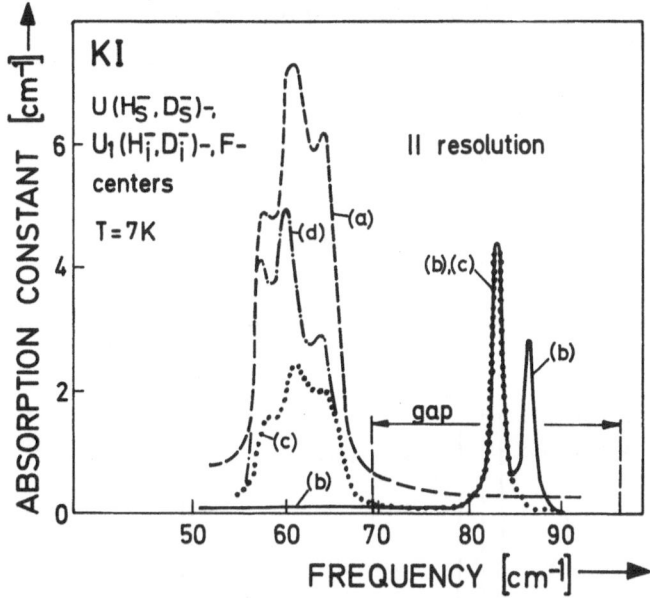

Fig. 39. Far infrared spectra at 7 K. (a) H_s^- centers (4×10^{18} cm^{-3}). (d) D_s^- centers. (b) after mercury arc irradiation (U_1, α, and F centers present), (c) after thermal recombination of α and U_1 centers (after Ref. [71])

can be given, since three new degrees of freedom are added to the section of the lattice which contains the defect. It is known, however, that the interstitial ion introduces a distortion of its cage of nearest neighbors [61]. It is proposed that this favors collective motions of the whole body of the interstitial ion and its four anion neighbors; a resulting resonant or localized mode displacement is shown in Fig. 41.

In Section 3.2 it was mentioned that infrared active modes of T_2 symmetry can couple to the localized mode. In Figs. 40 and 41 near infrared and far infrared spectra are compared. It is evident that the frequency of the strongest line in the sideband spectrum of the high frequency localized mode agrees quite well with the position of the line in the far infrared. This leads to the conclusion that the strong sideband peak is caused by a phonon, which also can be directly excited in the far infrared; i.e. one can make the correlation (see Section 3.2)

$$\Gamma = T_2^P = T_2(2)$$

This means that one can analyze the sideband spectrum of the high frequency localized mode with respect to T_2 peaks by investigating the far infrared spectra. Thereby, direct information on the coupling co-

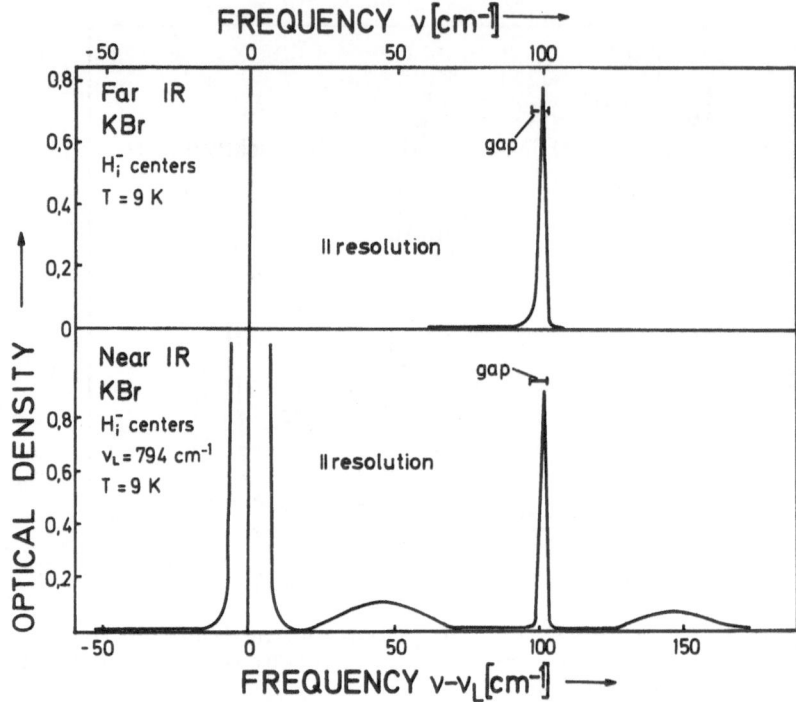

Fig. 40. Far infrared and near infrared vibrational absorption due to H_i^- centers in KBr (after Ref. [72])

efficient $A(\Gamma)$ is obtained. Using Eq. (28) and Eq. (23), employed for T_d symmetry, one finds

$$A(T_2(2)) \approx K \sum_\Gamma A(\Gamma)$$

where $K = 0.2$ and 0.5 for KI and KBr, respectively. If the integral absorption of the second line in the doublet in the KI spectrum is included, one would obtain $K \approx 0.4$. This would mean that the threefold degenerate $T_2(2)$ mode splits when it couples to the $T_2(1)$ localized mode. Whether this is the case or not could be proved e.g. by uniaxial stress experiments. The result that the coupling of the $T_2(2)$ mode is much stronger than the coupling of other modes is quite different from the case of the high frequency localized mode due to U centers for which coupling to A_{1g} modes is about three times stronger than coupling to modes of E_g and T_{2g} symmetry.

Another interesting effect is the disagreement of the precise far infrared and sideband peak positions (102 cm^{-1}, and 98.7 cm^{-1} for

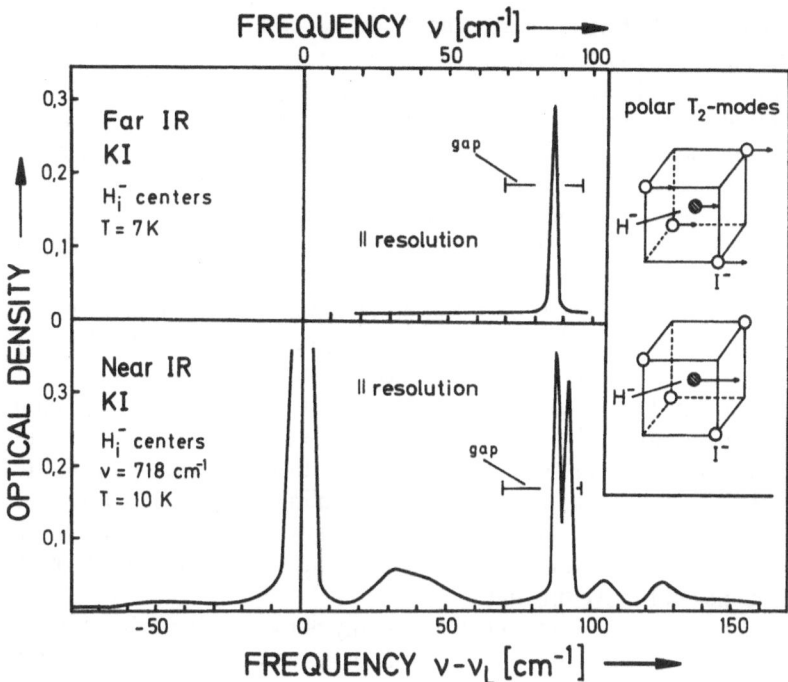

Fig. 41. Same as Fig. 40 but for KI. Insert shows model for vibrations $T_2(2)$, and $T_2(1)$ respectively

KBr, respectively). This disagreement is due to the frequency dependence of the anharmonic shift $\Delta_L(\omega)$ (see Section 3.1.2). This shifts the sideband by an amount which corresponds neither to the "free" nor to the "damped" far-infrared phonon frequency. This effect has been quantitatively treated for the U center high frequency localized mode [59].

Model calculations on the infrared vibrational absorption of U_1 centers in alkali halides were not yet performed. They could be done in analogy to calculations on the infrared vibrational absorption due to substitutional point defects by using the molecular Green function method.

5. Raman Spectra

All applications of Raman scattering to crystal lattices with point defects have taken place within the last 10 years. The reason is that high power lasers are necessary and the concentrations of the defect usually under investigation are low. In cubic crystals like alkali halides with point defects, Raman scattering is an especially powerful tool. The reason is that first order scattering is forbidden in the perfect lattice (see Section 2.2), i.e. any observed first order spectra are due only to the addition of the defect. When the defect is at a site of inversion symmetry, the only lattice modes coupled to the (virtual) electronic transition are of even parity. This means that in cubic crystals Raman scattering experiments yield information on that part of the phonon spectrum which cannot be directly excited in the infrared. Therefore, the Raman spectra complement the infrared absorption spectra.

Up to now, most Raman scattering experiments were performed in such a way that the energy of the exciting light was far from the electronic dipole transition (off-resonant or non-resonant Raman scattering). In this case one has to deal with a virtual electronic transition corresponding to a polarization of the electron shell. Since the electronic polarization consists of a mixing of the ground state with the optical-active excited states, the vibrational modes and the coupling coefficients involved are the same as those assisting the absorption process (see Section 2.2). The polarizability is assumed to be a constant, and independent of the frequency of the exciting light.

Special features occur when the energy of the exciting light, coincides with the electronic transition (resonant Raman scattering) [101, 102]. In this case the electronic polarizability can no longer be treated as a constant but may become very large and dependent on the frequency of the exciting light. The most characteristic feature of resonant Raman scattering for point defects in cubic crystals is the appearance of multiphonon spectra.

Section 5.1. reviews the experimental results on the non-resonant Raman scattering from U centers in the alkali halides KBr and KI [76], and in alkaline earth fluorides [66, 77]. Theoretical investigations were performed for U centers in alkali halides [24, 25]. Raman scattering from substitutional hydrogen centers in LiD was recently reported [103].

Near-resonant and resonant Raman scattering was observed for F centers in various alkali halides [102, 104, 105]. The present theoretical studies are based on off-resonance selection rules [27, 94, 105, 106]. A survey on these investigations is given in Section 5.2. Experimental results for F_A(Li) centers in KCl are mentioned in Section 5.3.

5.1. U Centers (H_s^-, D_s^- Centers)

In Section 3.1.4 it was found from symmetry arguments that the only phonon modes which can couple to the high frequency localized mode are of A_{1g}, E_g, and T_{2g} symmetry. These modes can couple to the electronic transition of the U center as well and should be observable in Raman scattering. In the most recent experiments by Montgomery et al. [76], however, only a weak scattering peak of A_{1g} symmetry was observed in KI with U centers. This line was found in the acoustic optical phonon gap in agreement with earlier observations in the sideband spectra [37]. In addition, well resolved second harmonics of the infrared active localized mode were found. The results for KBr: H_s^- and KI: H_s^- are shown in Figs. 42 and 43. In each spectrum three lines occur. These lines correspond to transitions from the vibrational ground state ($n = 0$) to the threefold split sixfold degenerate second excited state ($n = 2$) [see Fig. 6(a)].

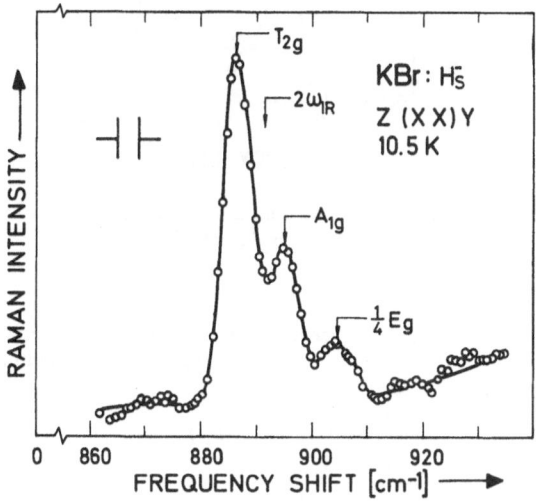

Fig. 42. $z(xx)y$ Raman spectrum of KBr : H_s^- in second harmonic region. $x \parallel [110]$, $y \parallel [\bar{1}10]$, and $z \parallel [001]$ (after Ref. [76])

The measured Raman frequencies may serve to calculate the coefficients C_1, C_2, and Ω of Eq. (15) ($B = 0$ for O_h symmetry). Their values are included in Table 5. They allow one to calculate the complete energy level scheme for the U center oscillator [see Fig. 6(a)]. As examples, Table 15 shows the calculated fundamental transition frequencies ($n = 0$ to $n = 1$) for KBr and KI with U centers. The good agreement with the experimentally determined infrared line positions (listed in parantheses) again indicates that the static well approximation, on which Eq. (15) is based, is a satisfactory approach for the U center localized oscillator, and yields a consistent description of infrared and Raman data.

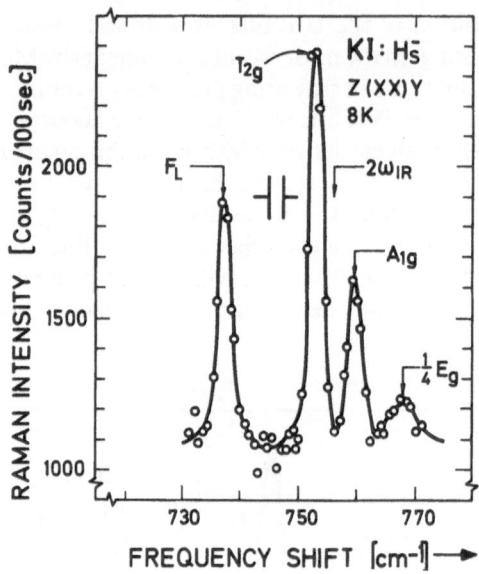

Fig. 43. $z(xx)y$ Raman spectrum of KI : H_s^- in second harmonic region. $x \parallel [110]$, $y \parallel [\bar{1}10]$, and $z \parallel [001]$. F_L denotes a plasma line of the laser (after Ref. [76])

Table 15. Calculated main line frequency for U center localized mode in KBr and KI (after Ref. [76]). Values determined by infrared absorption are in parantheses

Substance	Frequency $v_L(H_s^-)$ [cm^{-1}]	$v_L(D_s^-)$ [cm^{-1}]	Temperature [K]
KBr	446.5 (446)	319.1 (319)	11 (90)
KI	379 (382)	273 (280)	8 (90)

Table 16 lists the absolute Raman-scattering efficiencies and cross sections, which have been determined for the impurity induced modes in KI : H_s^-. It appears that the first order Raman scattering for the A_{1g} in-gap mode and the two-quantum scattering for the three harmonics is of comparable strength. As a possible explanation Montgomery et al. mention that the H_s^- ion might be a too weak electronic perturbation to produce a large first-order polarizability change. The relative intensities of the second harmonic peaks are not yet understood. They are in

Table 16. Absolute Raman efficiencies and differential cross sections for H_s^- centers in KI for 4880-Å exciting light (after Ref. [76])

| | Second harmonics | | | In-gap mode |
	T_{2g}	A_{1g}	E_g	(A_{1g})
Raman efficiency [$\times 10^{-12}$ cm^{-1} sr^{-1}]	2.23	0.906	1.33	1.11
Differential cross Section [$\times 10^{-30}$ cm^2/sr^{-1}]	3.33	1.35	1.99	1.66

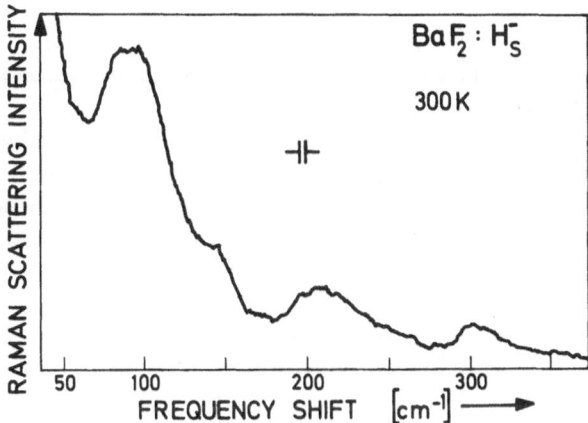

Fig. 44. High frequency sideband structure of the localized mode fundamental in BaF$_2$: H_s^- ($9.1 \cdot 10^{19}$ cm^{-3}) at 300 K as observed by Raman scattering. The scattering geometry displays T_2 components of the Raman tensor only. The spectrum was recorded with a time constant of 200 sec (after Ref. [66])

contradiction to the calculations which predict a much stronger scattering from $A_{1g} + E_g$ modes than from the T_{2g} mode.

An accurate Raman spectrum of localized mode sidebands was obtained in BaF$_2$: H_s^- by Harrington et al. [66]. Because in alkaline earth fluorides the site symmetry of U centers is T_d (see Section 3.1.1), the Raman tensor has components of A_1, E, and T_2 symmetry. The 300 K T_2 Raman spectrum of the sidebands is shown in Fig. 44; it is almost identical to the infrared spectrum of Fig. 12. This result is not surprising, since the dipole operator also transforms according to T_2 in T_d symmetry (see Section 2.2). Sideband scattering of A_1 and E symmetry was too weak to be observed in these experiments.

5.2. *F* Centers

Resonance enhancement, which occurs when irradiating near the electronic *F* band, allows one to obtain large Raman cross sections, even with relatively low concentrations of *F* centers (some 10^{17} to 10^{18} *F* centers cm^{-3}). This has been demonstrated experimentally by Worlock and Porto [104], Buchenauer et al. [105] and by Fitchen and Buchenauer [102] in a series of crystals like NaCl, NaBr, KF, KCl, and RbF with *F* centers. The *F* center induced Raman spectra are expected to be due to those lattice vibrations which interact with the excited electronic state. Assuming that the electron-phonon interaction is linear, and that the scattering center has octahedral symmetry, Henry and Slichter [108] have shown that the near resonance first-order Raman spectrum contains only modes of A_{1g}, E_g, and T_{2g} symmetry (see Section 2.2). These are modes which determine the width of the electronic *F* band.

The *F* center induced Raman spectra for NaCl, measured by Worlock and Porto, are shown in Fig. 45. The figure includes theoretical results for the first-order Raman spectra obtained by Benedek and Mulazzi [94] for the model of Fig. 4. For A_{01} and A_{14} the fitted values were taken (see Section 4.2). The i_{ij} are components of the first order Raman tensor given in Section 2.4. Because for NaCl $\beta \ll \alpha$ one has $i_{11} \approx i_{12}$.

From the comparison of experimental and theoretical data some interesting statements can be made: Firstly, discrepancies $\Delta\Omega$, between main peaks of observed and calculated curves occur. These were interpreted as being due to anharmonic contributions. They arise mainly from strong asymmetry in the potential of 1 nn of the *F* center. Secondly, the experimental data continue beyond the one phonon density of states, i.e. multiphonon scattering can not be neglected.

Even stronger multiphonon spectra occur for resonant Raman scattering. Experimentally this was shown by Fitchen and Buchenauer [102]. Theoretically Rebane et al. [109] found that the total *m*-phonon Raman cross sections follow the relation

$$(d\sigma/d\Omega)_m \propto (\hbar\omega_i)^4 |\phi_m(z)|^2 \tag{33}$$

where

$$\phi_m(z) = (2/m!\,2^m)^{1/2}\,(d/dz)^m \left[(\pi/2)^{1/2} + i \int_0^z e^{x^2}\,dx \right] e^{-z^2}$$

and

$$z = \hbar(\omega_i - \omega_F)/\sqrt{2}\,\sigma.$$

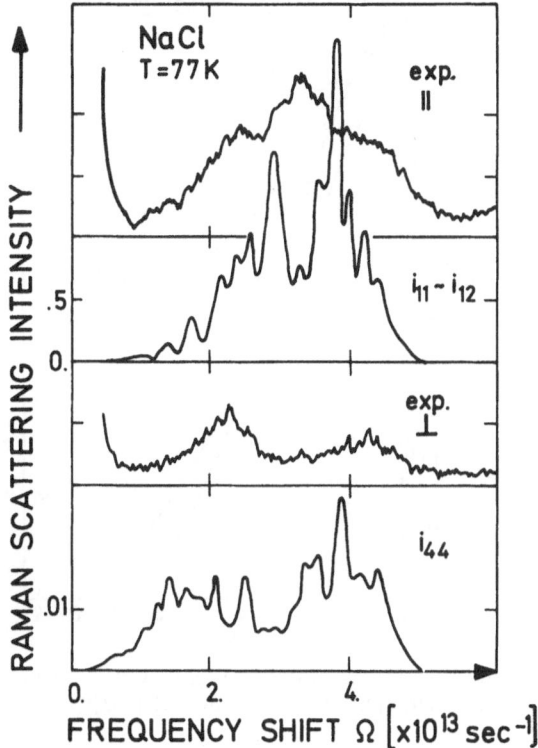

Fig. 45. Raman scattering intensity versus frequency shift for the F center in NaCl. The experimental data and the theoretical results, which were obtained by using fitted force constants for A_{01} and A_{14}, are compared (after Ref. [94])

σ is the second central moment of the band. z is the separation of the exciting (laser) frequency, ω_i, from the absorption band frequency, ω_F (in our case the center frequency of the electronic F band), in terms of the halfwidth of the band. The variation of $\phi_m(z)$ with z is shown in Fig. 46. This figure shows two remarkable features: first, the enhancement of scattered intensity as one approaches resonance, and second, the fact that multiphonon scattering intensities fall off slowly at resonance, but rapidly far from resonance. For comparison the assumed Gaussian shape of the absorption band is shown by the dashed curve.

The role of multiphonon scattering can be seen for the case of NaBr with F centers from Fig. 47. The most remarkable feature is a strong resonance peak at $136\,\mathrm{cm}^{-1}$ which occurs just above the phonon gap which extends from $105\,\mathrm{cm}^{-1}$ to $127\,\mathrm{cm}^{-1}$. Harmonics of this peak can be seen up to fourth order, and are indicated by the arrows. They can be

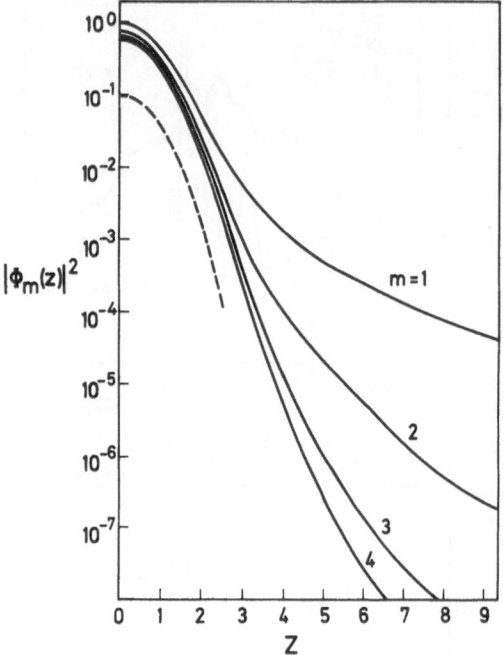

Fig. 46. Variation of the mth-order scattered intensity with separation of the incident frequency from the peak absorption frequency (after Ref. [102])

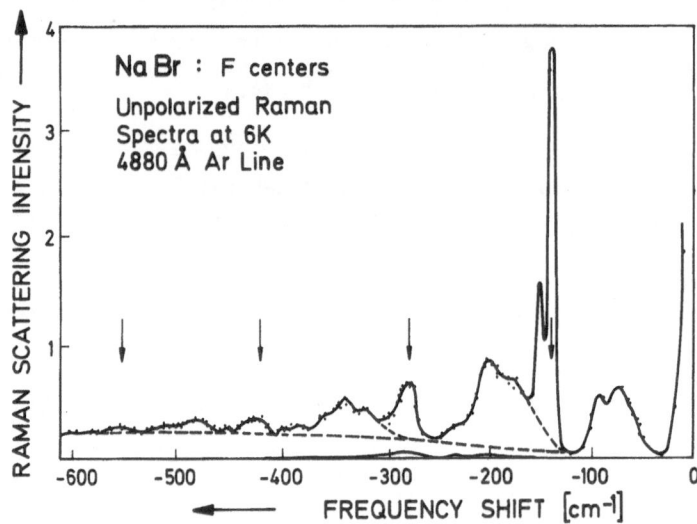

Fig. 47. Unpolarizes Stokes Raman spectrum at 6 K for F centers in NaBr with 4880Å excitation. The dashed lines suggest separation into different orders. The solid curve at the bottom is 2nd order scattering by pure crystal (after Ref. [102])

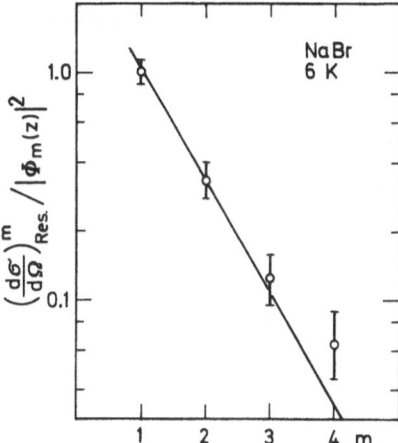

Fig. 48. Predicted and measured behavior of $(\sigma_i/\sigma_{\text{total}})^m$ for the $136\,\text{cm}^{-1}$ A_{1g} peak in NaBr with F centers (after Ref. [102])

used for testing Eq. (33) if it is generalized to the case of several coupled modes. This involves an additional factor $(\sigma_\Gamma/\sigma_{\text{Total}})^m$, which represents the relative scattering cross section for mode Γ. This factor can be determined independently from the integrated intensity of the peak and the total intensity in first order in different polarizations. Figure 48 shows the observed intensities up to fourth order. The straight line indicates the experimentally determined factor $(0.32)^m$.

A theoretical analysis of the NaBr Raman data was performed by Benedek and Mulazzi [106]. Off resonance selection rules were employed. The extended model was used and second order processes were taken into account. Figure 49 shows the results. The projected phonon density $\hat{\varrho}(A_{1g})$ and its convolution $\hat{\varrho}(A_{1g}) \times \hat{\varrho}(A_{1g})$ give the main contributions to the first and second order parallel spectra, respectively. The peak in the $\hat{\varrho}(A_{1g}) \times \hat{\varrho}(A_{1g})$ term is just the overtone of the $136\,\text{cm}^{-1}$ line. For $\perp$ spectra the terms $\hat{\varrho}(T_{2g})$ and $\hat{\varrho}(A_{1g}) \times \hat{\varrho}(T_{2g})$ are shown. Both, parallel and perpendicular Raman spectra are reproduced quite well. The superposition of the projected one and two-phonon densities was chosen in a way to fit the experimental data as good as possible. The coefficients of coupling for the electronic transition to lattice vibrations can be estimated from weighting factors. Unfortunately experimental stress data for NaBr with F centers are not yet available for comparison. The fitted force constants are $A01 = -0.67$, and $A14 = -0.25$.

The occurrence of the resonant mode in NaBr with F centers is in analogy to the results in Section 4.2.1 and 4.2.4. Its origin is due to large

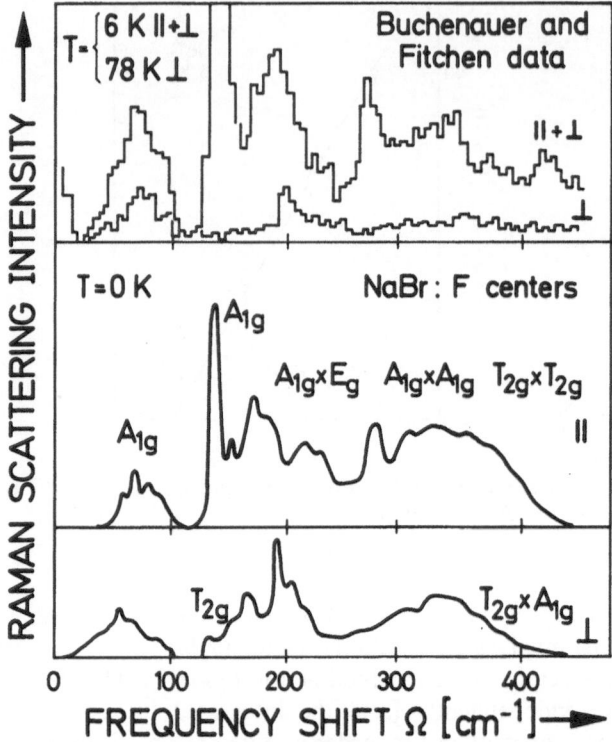

Fig. 49. Raman scattering intensity versus frequency shift for the F center in NaBr. The experimental data and the theoretical results, which were obtained by using fitted force constants for $A01$ and $A14$, are compared (after Ref. [106])

changes in local force constants. In fact, a strong gap mode of E_g symmetry in the Raman spectra of KI with F centers is predicted as well [94]. This is not yet measured.

New features occur in the resonant Raman spectra of KCl with F centers (see Fig. 50). A considerable change in scattered intensity is found when the wavelength of the excitation is varied from the F band to the K band. A broad band at about 190 cm^{-1} dominates the spectra for K band excitation. It is suggested that the 190 cm^{-1} band arises from long wavelength LO phonons. Since it is known that the excited states of the F center extend much further than the 1 nn shell, a coupling of LO phonons through the Fröhlich interaction has long been expected. K band states have radii of tens of Angströms and more. The total Raman cross section for one phonon scattering at 5145 Å is about $3 \cdot 10^{-22} \text{ cm}^2/\text{sr}$ (compare Table 16)!

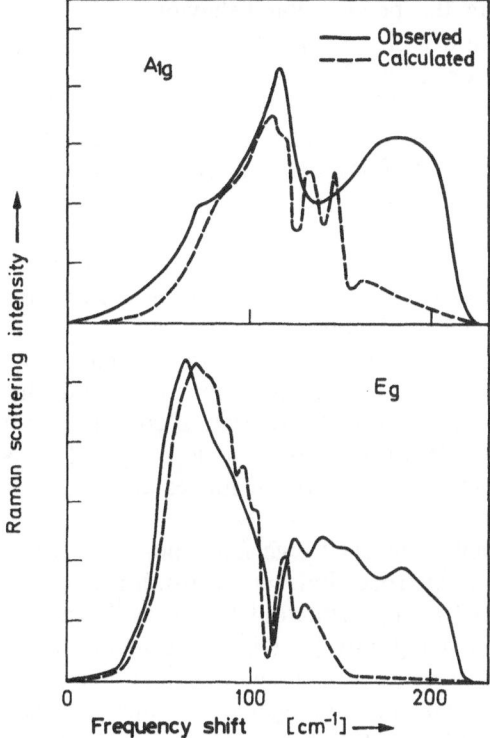

Fig. 50. Comparison of projected 1 nn phonon densities of states for perturbed modes in KCl ($A01 = -0.60$) with measured first order spectral densities for F centers in KCl at 6 K (after Ref. [110])

The calculated curves in Fig. 50 are obtained with a dynamical model based on the breathing shell model. Changes in force constant A_{01} were taken into account only. The best fit of the experimental curve outside the 190 cm^{-1} band was obtained for $A01 = -0.60$.

The relative strength of A_{1g}, E_g, and T_{2g} phonons was found to be 0.36 ± 0.05, 0.25 ± 0.04, and 0.40 ± 0.06. These values are in reasonable agreement with the relative contributions of these modes to the half width of the electronic F band as determined from stress induced dichroism.

5.3. F_A(Li) Centers

The resonant Raman scattering from F_A(Li) centers in KCl was studied by Fritz [107]. Besides changes of the spectrum in the F center scattering region, two new peaks at 288 cm^{-1} and 228 cm^{-1} (8 K) were observed. On substituting Li^7 instead of Li^6 the high-frequency line is shifted to

$268 \, \mathrm{cm}^{-1}$. From the polarization behavior of the scattered light for different orientations of the aligned center system, Fritz suggested that the Li^+ ion occupies an off-axis position. The fact that an overall C_{4v} symmetry has been established for the F_A center by ENDOR experiments [111] leads to the restriction that the off-axis position of the Li^+ ion must be at least fourfold degenerate about the center axis.

6. Summary

The present work shall serve to give a survey on the vibrational properties of electron and hydrogen centers in ionic crystals.

Electron and hydrogen centers have the common property that their mass is very small when compared with the mass of the host lattice ions. As a consequence, some of these centers give rise to high frequency localized modes. However, defect induced optical activity of lattice band modes, and acoustic resonant modes and in-gap modes are observed as well.

The most well established high frequency localized modes in ionic crystals are due to hydrogen ions on substitutional $[U(H_s^-, D_s^-) \text{ centers}]$ or interstitial $[U_1(H_i^-, D_i^-) \text{ centers}]$ lattice sites. Localized modes due to interstitial hydrogen atoms $[U_2(H_i^0, D_i^0) \text{ centers}]$ are fairly well established. They were found in alkali halides and in alkaline earth fluorides, and were mainly investigated from infrared absorption spectra and/or from Raman scattering experiments. For a number of crystals localized modes due to hydrogen ions and/or hydrogen atoms were established indirectly from vibronic spectra. They are not discussed in this work.

In alkali halides and alkaline earth fluorides the near infrared absorption spectra of U centers, and U_1 centers consist of a strong main line (fundamental) and sidebands. The main line is due to a vibrational transition from a non-degenerate ground state to a threefold degenerate first excited state. Higher harmonic vibrational transitions were observed for the case of U centers in the infrared absorption and/or Raman spectra. Their energy and splitting can be explained by an anharmonic oscillator, which is vibrating in a static potential well. The anharmonic coefficients, and thereby the complete oscillator level scheme can be calculated from the observed transition frequencies. The temperature dependent halfwidth and frequency shift of the fundamental transition, and the sideband structure observed, can be explained by anharmonic coupling of the localized mode to lattice band modes. Microscopic interaction mechanisms are suggested, and the amount of the coupling coefficients can be estimated. Experiments with uniaxial stress, electric fields, and mixed crystals support and complement the results.

In the far infrared spectra acoustic resonant modes and in-gap modes were found for U centers, F'centers, and for F centers, $F_A(Na)$ centers, U_1 centers, respectively. Resonant modes eventually present within the optical band can not be directly observed in ionic crystals because of the strong "reststrahl-absorption".

Raman scattering experiments from F centers in alkali halides did yield information on the density of even parity lattice modes and on their relative amount of coupling to the electronic transition.

From the number of vibrational transitions observed in the infrared and/or Raman spectra the local site symmetry of a defect can be determined. Its local force constants and vibrational amplitudes can be derived from the frequency positions of lines, and from their relative intensities, respectively. From the local force constants local static relaxations can be estimated. Because these relaxations affect the overlap between the electronic wave function of the defect and its nearest neighbors, they are important in a calculation of the electronic energy of a defect.

From the defect induced band mode absorption, and from localized mode sidebands, information on host lattice phonons, their densities, interactions, and perturbations is obtained.

7. Appendix

Table A1. Highest band and gap edge frequencies for alkali halides[a]

Substance	ω_{max} [cm^{-1}]	Gap edge freq. lower	upper	T [K]	References
LiF	657	–	–		[112]
NaF	427	–	–	295	[124]
NaCl	261	–	–	80	[113]
NaBr	207	105	127	295	[114]
NaI	170	77	117	100	[115]
KCl	212	–	–	80	[116]
KBr	167	94	102	90	[115]
KI	142	69	98	95	[117]
RbF	283	125	163	80	[118]
RbCl	173	100	113	80	[118]
RbBr	130	–	–	80	[119]
RbI	105	–	–	80	[120]
CsCl	163	86	92	78	[121]
CsBr	112	–	–	80	[122]
CsI	84	–	–	295	[123]

[a] Values derived from Inelastic Neutron Scattering Experiments.

158   *D. Bäuerle:*

Acknowledgements. I am indebted to P. Grosse for his encouragement in writing this review, and would like to further thank him, together with H. Bilz, L. Genzel, R. Hayes, T. P. Martin, H. Pick, and R. Zeyher, for several stimulating discussions. I am also grateful to D. B. Fitchen for preprints of his recent papers and for allowing me to incorporate them in this review.

References

1. Fowler, W. B.: Color Centers in Alkali Halides. New York: Academic Press 1968.
2. Pick, H.: Optical Properties of Solids. Ed. Abelès, F. Amsterdam: North-Holland Publ. Comp. 1972.
3. Pick, H.: Struktur von Störstellen in Alkalihalogenidkristallen. Springer Tracts, Vol. 38. Berlin-Heidelberg-New York: Springer 1965.
4. Schulman, J. H., Compton, W. D.: Color Centers in Solids. New York: Pergamon Press 1963.
5. Elliott, R. J., Hayes, W., Jones, G. D., Macdonald, H. F., Sennett, C. T.: Proc. Roy. Soc. **289**, 1 (1965).
6. Bessent, R. G., Hayes, W., Hodby, I. W.: Phys. Letters **15**, 115 (1965).
7. Jones, G. D., Peled, S., Rosenwaks, S., Yatsiv, S.: Phys. Rev. **183**, 353 (1969).
8. Loudon, R.: Proc. Phys. Soc. (London) **84**, 379 (1964).
9. Loudon, R.: Advan. Phys. **13**, 432 (1964).
10. Knox, R. S., Gold, A.: Symmetry in the Solid State. New York Benjamin 1964.
11. Tinkham, M.: Group Theory and Quantum Mechanics. New York: MacGraw Hill 1964.
12. Mulliken, R. S.: J. Chem. Phys. **3**, 375 (1935).
13. Bouckaert, L. P., Smoluchowski, R., Wigner, E.: Phys. Rev. **50**, 58 (1936).
14. Bethe, H. A.: Ann. Phys. **3**, 133 (1929).
15. von der Lage, F. C., Bethe, H. A.: Phys. Rev. **71**, 619 (1947).
16. Howarth, D. J., Jones, H.: Proc. Phys. Soc. (London) A **65** 353 (1952).
17. Genzel, L., Festkörperprobleme, Vol., VI. London: Pergamon 1966.
18. Sievers, A. J.: Elementary Excitations in Solids, p. 193. New York: Plenum Press 1969.
19. Benedek, G., Nardelli, G. F.: Phys. Rev. **155**, 1004 (1967).
20. Klein, M. V.: see Ref. [1].
21. Hübner, R.: Z. Physik **222**, 380 (1969).
22. Kühner, D., Wagner, M.: Z. Physik **207**, 111, (1968).
23. Gethins, T., Timusk, T., Woll, E. J.: Phys. Rev. **157**, 744 (1967).
24. Sennet, C. T.: J. Phys. Chem. Solids **26**, 1097 (1965).
25. Xinh, N. X., Maradudin, A. A., Coldwell-Horsfall, R. A.: J. Phys. Radium **26**, 717 (1965).
26. Maradudin, A. A.: Solid State Phys., Vol. 19. Ed. Seitz, F., and Turnbull, D. New York: Academic Press 1966.
27. Benedek, G., Nardelli, G. F.: Phys. Rev. **154**, 872 (1967).
28. Gebhardt, W., Maier, K.: Phys. Stat. Sol. **8**, 303 (1965).
29. Rosenstock, H., Klick, C. C.: Phys. Rev. **119**, 1198 (1960).
30. Schäfer, G.: J. Phys. Chem. Sol. **12**, 233 (1960).
31. Price, W. C., Wilkinson, G. R.: Final Technical Report Nr. 2 (dec. 1960), US Army Contract No DA–91–591–EUC 1908, OI–4201–60 (R. + D. 260).
32. Mitsuishi, A., Yoshinaga, H.: Progr. Theoret. Phys. (Kyoto) Suppl. **23**, 241 (1962).
33. Mirlin, D. N., Reshina, I., I.: Sov. Phys. Sol. State **6**, 728 (1964).
34. Mirlin, D. N., Reshina, I. I.: Sov. Phys. Sol. State **6**, 2454 (1965).
35. Mirlin, D. N., Reshina, N. N.: Sov. Phys. Solid State **8**, 116 (1966).
36. Mitra, S. S., Brada, Y.: Phys. Letters **17**, 19 (1965).
37. Fritz, B., Groß, U., Bäuerle, D.: Phys. Stat. Sol. **11**, 231 (1965).
38. Dötsch, H., Gebhardt, W., Martius, C. H.: Phys. Stat. Sol. **3**, 9 (1965).

39. Mitra, S. S., Singh, R. R.: Phys. Rev. Letters 16, 694 (1966).
40. Timusk, T., Klein, M. V.: Phys. Rev. 141, 664 (1966).
41. Bäuerle, D., Fritz, B.: Phys. Stat. Sol. 24, 207 (1967).
42. Barth, W., Fritz, B.: Phys. Stat. Sol. 19, 515 (1967).
43. Fritz, B., Gerlach, J., Groß, U.: Localized Excitations in Solids. Ed. Wallis, R. F. New York: Plenum Press 1968.
44. Dötsch, H.: Phys. Stat. Sol. 31, 649 (1969).
45. Dötsch, H., Mitra, S. S.: Phys. Rev. 178, 1492 (1969).
46. MacPherson, R. W., Timusk, T.: Can. J. Phys. 48, 2146 (1970).
47. de Souza, M., Gongora, A. D., Aegerter, M., Lüty, F.: Phys. Rev. Letters 25, 1426 (1970).
48. Kurz, G., Susman, S., Internat. Conf. "Color Centers in Ionic Crystals", Reading, U. K., 1971.
49. Olson, C. G., Lynch, D. W.: Phys. Rev. 4, 1990, (1971).
50. Wallis, R., Maraduin, A. A.: Progr. Theor. Phys. (Kyoto) 24, 1055 (1960).
51. Bilz, H., Strauch, D., Fritz, B.: J. Phys. (Paris) 21. Suppl. C 2 – 3 (1966).
52. Bilz, H., Zeyher, R., Wehner, R. K.: Phys. Stat. Sol. 20, K 167 (1967).
53. Page, J. B., Dick, B. G.: Phys. Rev. 163, 910 (1967).
54. Xinh, N. X.: Phys. Rev. 163, 896 (1967).
55. Page, J. B., Strauch, D.: Phys. Stat. Sol. 24, 469 (1967).
56. Strauch, D., Page, J. B.: Phys. Stat. Sol. 30, 495 (1968).
57. Cunningham, S. L., Hardy, J. R.: Solid State Commun. 6, 769 (1968).
58. Strauch, D.: Phys. Stat. Sol. 30, 495 (1968).
59. Zeyher, R., Bilz, H., see Ref. [43], p. 767.
60. Zeyher, R., Bilz, H.: Phys. Stat. Sol. 31, 157 (1969).
61. Striefler, M., Jaswal, S. S.: J. Phys. Chem. Solids 30, 827 (1969).
62. Haridasan, T. M., Krishnamurthy, N.: J. Indian Inst. Sci. 51, 1 (1969).
63. Boese, F. K., Wagner, M.: Z. Physik 235, 140 (1970).
64. Lannoo, M., Dobrzynski, L.: J. Phys. Chem. Solids 33, 1447 (1972).
65. Harrington, J. A., Walker, C. T.: Phys. Stat. Sol. (b) 43, 619 (1971).
66. Harrington, J. A., Harley, R. T., Walker, C. T.: Solid State Commun. 9, 683 (1971).
67. Harrington, J. A., Weber, R.: Solid State Commun. 11, 1435 (1972).
68. Fritz, B., see Ref. [43].
69. Fritz, B.: J. Phys. Chem. Sol. 23, 375 (1962).
70. Gross, U., Bron, W. E.: Phys. Letters 25 A, 312 (1967).
71. Bäuerle, D., Fritz, B.: Phys. Stat. Sol. 29, 639 (1968).
72. Dürr, U., Bäuerle, D.: Z. Physik 233, 94 (1970).
73. Akhvlediani, Z. G., Politov, N. G.: ZHETF, 10, 249 (1969).
74. Politov, N. G., Akhvlediani, Z. G., see Ref. [48].
75. Shamu, R. E., Hartmann, W. M., Yasaitis, E. L.: Phys. Rev. 170, 822 (1968).
76. Montgomery, G. P., Fenner, W. R., Klein, M. V., Timusk, T.: Phys. Rev. B 5, 3343 (1972).
77. Harrington, J. A., Harley, R. T., Walker, C. T.: Solid State Comm. 8, 407 (1970).
78. Hayes, W., Macdonald, H. F.: Proc. Roy. Soc. A 297, 503 (1967).
79. Jones, G. D., Satten, R. A.: Phys. Rev. 147, 566 (1966).
80. Dötsch, H.: Ph. D. Thesis, University of Frankfurt, 1967.
81. Lee, L. C., Faust, W. L.: Phys. Rev. Letters 26, 648 (1971).
82. Hayes, W., Macdonald, H. F., Elliott, R.: J. Phys. Rev. Letters 15, 961 (1965).
83. Cowley, R. A.: Advan. Phys. 12, 421 (1963).
84. Hurrell, J. P., Minkiewiez, V. J.: Solid State Commun. 8, 463 (1970).
85. Benedek, G., Nardelli, G. F.: Phys. Rev. Letters 16, 517 (1966).
86. Bäuerle, D.: Ph. D. Thesis, University of Stuttgart, 1969.
87. Bäuerle, D., Hübner, R.: Phys. Rev. B 2, 4252 (1970).
88. de Souza, M. F. Lüty, F.: see Ref. [48].
89. Woll, E. J., Gethins, T., Timusk, T.: Can. J. Phys. 46, 2263 (1968).

90. Bäuerle, D.: Internat. Symposium "Color Centers in Alkali Halides", Rom, 1968.
91. Bäuerle, D., Fritz, B.: Solid State Commun. 6, 453 (1968).
92. Ram, P. N., Agrawal, B. K.: Solid State Commun. 11, 1719 (1972).
93. Markham, J. J.: Solid State Phys. Suppl. 8, ed. Seitz, F. and Turnbull, D. New York: Academic Press 1965.
94. Benedek, G., Mulazzi, E.: Phys. Rev. 179, 906 (1969).
95. Singh, R., Mitra, S. S.: Phys. Rev. B 2, 1070 (1970).
96. See e.g. Dexter, D. L.: Solid State Physics. Vol. 6, p. 353. ed. Seitz, F., Turnbull, D. New York: Academic Press 1958.
97. Jaswal, S. S.: Phys. Rev. 140, A 687 (1965).
98. Ward, R. W., Timusk, T.: Phys. Rev. B 5, 2331 (1972).
99. Kersten, R.: Ph. D. Thesis, University of Stuttgart, 1970.
100. Gnaedinger, R. J.: J. Chem. Phys. 21, 323 (1953).
101. Martin, R. M.: Light Scattering in Solids. Ed. Balkanski, M. Paris: Flamarion Sci. 1971.
102. Fitchen, D. B., Buchenauer, C. J.: Physics of Impurity Centers in Crystals, Tallin, 1972, p. 277.
103. Wolfram, G., Jaswal, S. S., Sharma, T. P.: Phys. Rev. Letters 29, 160 (1972).
104. Worlock, J. M., Porto, S. P. S.: Phys. Rev. Letters 15, 697 (1965).
105. Buchenauer, C. J., Fitchen, D. B., Page, J. B.: Light Scattering Spectra of Solids, p. 521. Ed. Wright, G. B. Berlin-Heidelberg-New York: Springer 1969.
106. Bendek, G., Mulazzi, E.: see Ref. [105], p. 531.
107. Fritz, B.: see Ref. [43], p. 496.
108. Henry, C. H., Slichter, C. P.: see Ref. [3], p. 351.
109. Rebane, K., Hizhnyakov, V., Tehver, I.: Izv. Akad. Nauk. Est. SSR, Ser. Fiz.-Mat. Tekh. Nauk. 16, 207 (1967).
110. Fitchen, D. B., Buchenauer, C. J.: see Ref. [48].
111. Mieher, R. L.: Phys. Rev. Letters 8, 362 (1962).
112. Dolling, G., Smith, H. G., Nicklow, R. M., Vijayaraghavan, P. R., Wilkinson, M. K.: Phys. Rev. 168, 970 (1968).
113. Raunio, G., Rolandson, S.: Phys. Rev. B 2, 2098 (1970).
114. Reid, J. S., Smith, T., Buyers, W. J. L.: Phys. Rev. B 1, 1833 (1970).
115. Cowley, R. A., Cochran, W., Brockhouse, B. N., Woods, A. D. B.: Phys. Rev. 119, 980 (1960).
116. Raunio, G., Almqvist, L.: Phys. Stat. Sol. 33, 209 (1969).
117. Dolling, G., Cowley, R. A., Schittenhelm, C., Thorson, I. M.: Phys. Rev. 147, 577 (1966).
118. Raunio, G., Rolandson, S.: J. Phys. C 3, 1013 (1970).
119. Rolandson, S., Raunio, G.: J. Phys. C 4, 958 (1971).
120. Raunio, G., Rolandson, S.: Phys. Stat. Sol. (b) 40, 749 (1970).
121. Ahmad, A. A. Z., Smith, H. G., Wakabayashi, N., Wilkinson, M. K.: Phys. Rev. B 6, 3956 (1972).
122. Rolandson, S., Raunio, G.: Phys. Rev. B 4, 4617 (1971).
123. Bührer, W., Hälg, W.: Phys. Stat. Sol. (b) 46, 679 (1971).
124. Buyers, W. J. L.: Phys. Rev. 153, 923 (1967).

Dr. D. Bäuerle
Max-Planck-Institut für Festkörperforschung
7000 Stuttgart 1
Heilbronner Str. 69
and
Philips Forschungslaboratorium
5100 Aachen
Weißhausstr. 4
Germany

Factor Group Analysis Revisited and Unified

J. BEHRINGER

Contents

Abstract

Factor group analysis under the condition $k = 0$ is reviewed and presented in compact formulation applicable to all crystal structures. In addition to the naive inspection method more advanced techniques are discussed.

1. Introduction

Factor group analysis (FGA) aims at the determination of the irreducible representations (irreps) of the various types of collective nuclear motions in ideal crystals. We are concerned here only with the <u>ordinary</u> FGA which applies to the case that the velocities of all nuclei in homologous positions at any instant exactly agree in magnitude and direction[1].

This condition for ordinary FGA can be described briefly by the equation

$$k = 0 \tag{1}$$

for the elastic wave vector k. ($k = 0$ corresponds to the center of the first Brillouin zone.) Condition (1) represents a strong constraint on crystal kinetics. It implies complete uniformity of the simultaneously occurring nuclear motions in all PCs and thereby reduces the infinite number of nuclear degrees of freedom of the free ideal crystal to only $3 N_A^{PC}$ where N_A^{PC} is the number of atoms within one PC. In order to know the locations and motions of the nuclei in the total crystal at some instant it is sufficient under condition (1) to know the nuclear locations and motions in only <u>one</u> PC at this instant. In brief, the whole crystal with its composition and motion by virtue of (1) appears simply as the spatial repetition (with the periodicity of the primitive translations) of one PC.

Ordinary FGA was initiated by Bhagavantam and Venkatarayudu [1] and discussed and further developed by many authors [2]. Its applicability to IR and Raman (R) spectroscopy depends on the compatibility of condition (1) with the physical laws governing IR absorption and R scattering in crystals. For the Raman effect (RE) the following remarks apply. The length of the phonon k-vectors in optically excited R spectra is 2 or 3 orders of magnitude smaller than the typical diameter of the first Brillouin zone. Therefore, because of energy and quasi-momentum conservation, the wave vectors k of the phonons created or annihilated by first order R interactions in crystals obey condition (1) to a very good approximation, although not exactly. In second and higher order REs the deviation of the phonon k-vectors from zero can be considerable. Therefore the ordinary FGA can be applied to first order R scattering with almost no restrictions whereas in higher order R scattering a FGA generalized for $k \neq 0$ is needed.

---

[1] Two positions are called h o m o l o g o u s when one is sent into the other by a primitive translation. Non-identical homologous points necessarily belong to different primitive cells. A p r i m i t i v e c e l l (PC) is always understood as a unit cell of smallest volume, particularly as the rhombohedral cell in the trigonal space groups R3, R$\bar{3}$, R32, R3m, R3c, R$\bar{3}$m, R$\bar{3}$c.

2. Terminology and Notation

2.1. Groups

$\mathcal{S}$ s p a c e   g r o u p

$\mathcal{P}$ d i r e c t i o n   g r o u p (or crystal class point group) belonging to $\mathcal{S}$
($\mathcal{P}$ from "point")[2]

$\mathcal{T}$ t r a n s l a t i o n   g r o u p (invariant subgroup of primitive translations)
contained in $\mathcal{S}$

$\mathcal{P}' \equiv \mathcal{S}/\mathcal{T}$ f a c t o r   g r o u p (isomorphic with, but conceptually different — therefore the prime — from $\mathcal{P}$)[3]

$\mathcal{L}_p$ s i t e   g r o u p belonging to point P in the crystal space ($\mathcal{L}$ from
German "Lage"), see Sect. 7.

2.2. Group Elements

$\{R|t\} = \{R|\tau_R + a_\alpha\} \in \mathcal{S}$ space group element, S e i t z   s p a c e
g r o u p   o p e r a t o r [3] (R rotational part, $t = \tau_R + a_\alpha$ translational part; τ_R n o n p r i m i t i v e   t r a n s l a t i o n belonging to R ;
$a_\alpha \equiv a_{\alpha_1 \alpha_2 \alpha_3} = \sum_{i=1}^{3} \alpha_i a_i$ p r i m i t i v e (or p u r e or l a t t i c e)
t r a n s l a t i o n, $\alpha_i, , i = 1, 2, 3$ scalar integers, a_i b a s i c   p r i m i -
t i v e   t r a n s l a t i o n s   o r   f u n d a m e n t a l   t r a n s l a t i o n   v e c t o r s
spanning the lattice PC)

R $\in \mathcal{P}$ direction group element (point group operator, equivalent to
$\{R|0\}$)[4]

$\{E|a_\alpha\} \in \mathcal{T}$ translation group element (primitive translation operator)

$R' \equiv \{R|\tau_R\}\mathcal{T} = \mathcal{T}\{R|\tau_R\} \in \mathcal{P}'$ factor group element (= coset of $\mathcal{T}$
in $\mathcal{S}$ with representative $\{R|\tau_R\}$)[5]

---

[2] $\mathcal{P}$ is found by simply cancelling the superscript in the Schoenflies
symbol for $\mathcal{S}$, e.g. for $\mathcal{S} = \mathfrak{D}_{3d}^6$ we have $\mathcal{P} = \mathfrak{D}_{3d}$.

[3] The isomorphism $\mathcal{P} \leftrightarrow \mathcal{P}'$ implies equality of orders $[\mathcal{P}] = [\mathcal{P}']$.

[4] Attention is drawn to the fact that the absolute orientation, i. e. the
orientation relative to the crystal space, of the symmetry elements (axes,
planes) associated with the elements $R \in \mathcal{P}$ is fixed.

[5] The $[\mathcal{P}]$ different space group elements $\{R|\tau_R\}$ which can be under-
stood as special representatives of the cosets R' of $\mathcal{T}$ in $\mathcal{S}$ in general do
not constitute a group because the group closure property may be absent
for this set. Only if $\tau_R = 0$ for all $R \in \mathcal{P}$ this set is a group. $\mathcal{S}$ is then
called s y m m o r p h i c . R' can be considered as a succession (in any or-
der) of the space group operation $\{R|\tau_R\}$ and an indefinite primitive trans-
lation $\{E|a_\alpha\}$.

2.3. Material Structure Units of the Crystal

In FGA two different approximations to the material content of a PC are used: the atomic and the molecular aspect. In the <u>atomic aspect</u> the matter contained in a PC is considered as a system of atoms, or, in fully sufficient simplification, of atomic nuclei. (The electrons explicitly do not play any role.) We therefore, when using this aspect, shall speak of a t o m s , symbolized by A, meaning atomic nuclei. In the <u>molecular</u> <u>aspect</u> the matter in a PC is thought of as a system of molecules of different sizes, shapes, and mutual arrangement. For the sake of unification and convenience we shall use the term "molecule" in a generalized sense: M o l e c u l e shall be understood as any atomic complex in the PC whatsoever (electrically charged (ion) or not and comprising any number of atoms — even only one in the limit). In other words, molecules are simply meant as the larger or smaller building blocks of the crystal which possess some enhanced internal stability within their surroundings. Of course the atomic aggregates which we consider as "molecules" should be chosen in a physically reasonable way, i. e. by taking account of the relative strength of the mutual binding forces. It goes without saying that this is not always possible without arbitrariness.

For the following argument it will be necessary to distinguish the molecules making up the crystal by the following geometrical classification of their equilibrium configurations:

(A) p o i n t - s h a p e d (i. e. monoatomic)
(B) l i n e a r (diatomic or polyatomic)
(C) n o n l i n e a r (polyatomic)

It will turn out later that in case B a further subdivision of the molecular types must be introduced if the space group $\mathcal{S}$ contains one or several two-fold rotations or/and reflections leaving the molecular centers of mass (COM) invariant (i. e. contained in the site group belonging to the respective COM): The linear molecules then should be divided into those having their symmetry axis ζ parallel and those having it perpendicular to the rotation axis (or axes) or/and reflection planes. (Elements of $\mathcal{P}$ associated with axes or planes oblique to ζ can be left out of consideration because they are not symmetry operations of the molecule. The influence of these elements will cancel out automatically in the following calculations.)

We shall denote the different species of atoms and molecules constituting the material content of a PC by the symbols A_s and M_s , respectively. The subscript s stands for "species" and is used to distinguish atoms by their chemical (and isotopic) character and molecules by their atomic composition and geometrical configuration, and linear molecules additionally by the orientation of their axis ζ relative to absolute crystal space. We shall specify the individual s by 1α, $1b$, 2α etc., the first part 1, 2, ... (sufficient for atoms and point-shaped and nonlinear molecules) distinguishing the chemical and physical properties and the second part $a, b, ...$ (needed in linear molecules only) the different orientations of the ζ-axis in space. E. g. in gypsum $CaSO_4 \cdot 2H_2O$ the following atoms A_s and molecules M_s can be assumed: $A_1 = Ca$, $A_2 = S$, $A_3 = O$, $A_4 = H$; $M_1 = Ca^{++}$, $M_2 = SO_4^{--}$, $M_3 = H_2O$.

2.4. Numbers of Particles in a PC

Let the crystal be built of atoms $A_{s'}$ of any species s' and molecules M_s of any species s. Then we introduce the count numbers

$N_{A_{s'}}^{M_s}(R')$ of <u>those</u> atoms $A_{s'}$ in one molecule M_s ,

$N_A^{M_s}(R') = \sum_{s'} N_{A_{s'}}^{M_s}(R')$ of the sum of <u>those</u> atoms A in one molecule M_s ,

$N_{M_s}^{PC}(R')$ of <u>those</u> molecules M_s in one PC,

$N_{A_{s'}}^{PC}(R') = \sum_s N_{A_{s'}}^{M_s}(R') N_{M_s}^{PC}(R')$ of <u>those</u> atoms $A_{s'}$ in one PC,

$N_A^{PC}(R') = \sum_s N_A^{M_s}(R') N_{M_s}^{PC}(R') = \sum_{ss'} N_{A_{s'}}^{M_s}(R') N_{M_s}^{PC}(R')$ of the sum of <u>those</u> atoms A in one PC,

$N_M^{PC}(R') = \sum_s N_{M_s}^{PC}(R')$ of the sum of <u>those</u> molecules M in one PC,

the CsOM of which remain invariant under application of the factor group element R' to the crystal, or, in other words, which apart from an eventual primitive translation $a_\alpha \neq 0$ do not change the COM positions when the space group operation $\{R | \tau_R\}$ is applied.

In the special case $R' = E' \equiv \mathcal{J}$ the COM positions of <u>all</u> particles are left invariant (apart from an indefinite pure translation) so that the above symbols give the <u>total</u> number of the particles denoted by the subscript in the larger unit denoted by the superscript. In this case the argument (E') in parentheses can be omitted.

The above symbols look somewhat overloaded with indices, but this prolixity will prove advantageous for a secure handling of the following formulae. In what follows essentially only $N_A^{PC}(R')$ and $N_{M_s}^{PC}(R')$ will be needed.

3. Representations of the Factor Group

Under condition (1) the following types of collective nuclear motions in
the crystal can be defined:

          1. Translations [6] (T)

          2. Optical vibrations

              a) External vibrations

                  α) Translatory vibrations (TV)

                  β) Rotatory (or torsional) vibrations or librations (RV)

              b) Internal vibrations (V)

Every one of these types of motion possesses a certain number of
degrees of freedom (to be calculated below, see Sect. 6a) and can be
described by the time-dependence of a normal coordinate vector $Q^{\cdots}$ with
a dimension equal to this number. All these normal coordinate vectors can
be combined by aligning their components Q_l in definite order ("external
product") to give the $3N_A^{PC}$-dimensional total normal coordinate
vector

$$Q = \{Q_1, \ldots, Q_l, \ldots, Q_{3N_A^{PC}}\} = \{Q^T, Q^{TV}, Q^{RV}, Q^V\}. \qquad (2)$$

The abstract space spanned by Q (normal coordinate space) and
its subspaces spanned by the parts Q^T, Q^{TV}, Q^{RV}, Q^V of Q can be considered as representation (rep) spaces for the factor group $\mathscr{P}'$. The corresponding (not necessarily irreducible) reps of $\mathscr{P}'$ will be designated by
$\mathscr{D}^T, \mathscr{D}^{TV}, \mathscr{D}^{RV}, \mathscr{D}^V, \mathscr{D}$, respectively. They obey the relation

$$\mathscr{D} = \mathscr{D}^T \oplus \mathscr{D}^{TV} \oplus \mathscr{D}^{RV} \oplus \mathscr{D}^V \qquad (3)$$

($\oplus$ "direct sum") which means that the rep matrices $D(R')$ of <u>all</u> elements $R' \in \mathscr{P}'$ in the rep $\mathscr{D}$ uniformly decompose in blocks along the
principal diagonal according to

$$D(R') = \begin{pmatrix} D^T(R') & & & \\ & D^{TV}(R') & & \\ & & D^{RV}(R') & \\ & & & D^V(R') \end{pmatrix}, \qquad (4)$$

all non-diagonal blocks containing only zeros.

For IR and R spectroscopy three more reps of $\mathscr{P}'$ are of importance:

---

[6] By condition (1) travelling waves are forbidden; the translations are
the limiting case for the acoustical vibrations.

1. $\mathcal{D}^{\mu}$ (dipole moment rep). The rep space is spanned by the 3 components of the electric dipole moment vector μ. $\mathcal{D}^{\mu}$ is equivalent to $\mathcal{D}^{T}$. This rep is important for IR spectroscopy.

2. $\mathcal{D}^{[\alpha]}$ (symmetric polarizability tensor rep). The rep space is spanned by the 6 components of the symmetric part $[\alpha]$ of the polarizability tensor α. This rep is important for R spectroscopy.

3. $\mathcal{D}^{\{\alpha\}}$ (antisymmetric polarizability tensor rep). The rep space is spanned by the 3 components (equivalent to the components of an axial vector) of the antisymmetric part $\{\alpha\}$ of the polarizability tensor α. This rep is needed in R spectroscopy only in certain special cases.

4. Characters

The determination of the irreps of the crystal motions discussed in Sect. 3 has to start from the characters of the reps corresponding to these motions. In Table 1 all information concerning these characters is collected in compact form applicable to all possible cases. The symbols not yet explained will be discussed in the following notes.

Table 1.

Types of crystal motion	Rep $\mathcal{D}^{\cdots}$	Character of $R' \in \mathcal{P}'$ $\chi^{\cdots}(R')$	Eq. No.
1. Translations	$\mathcal{D}^{T}$	$\chi^{T}(R') = f_R$	(5)
2. Optical vibrations			
a. External vibrations			
α. Translatory vibrations	$\mathcal{D}^{TV}$	$\chi^{TV}(R') = (N_M^{PC}(R')-1)f_R = (\sum_s N_{M_s}^{PC}(R')-1)f_R$	(6)
β. Rotatory vibrations	$\mathcal{D}^{RV}$	$\chi^{RV}(R') = \sum_s N_{M_s}^{PC}(R') f_R'^{(s)}$	(7)
b. Internal vibrations	$\mathcal{D}^{V}$	$\chi^{V}(R') = \chi(R') - \chi^{T}(R') - \chi^{TV}(R') - \chi^{RV}(R')$	(8)
3. Total of motions	$\mathcal{D}$	$\chi(R') = N_A^{PC}(R') f_R$	(9)
Electric dipole moment	$\mathcal{D}^{\mu}$	$\chi^{\mu}(R') = f_R$	(10)
Polarizability, symm. part	$\mathcal{D}^{[\alpha]}$	$\chi^{[\alpha]}(R') = f_R''$	(11)
" , antisymm. part	$\mathcal{D}^{\{\alpha\}}$	$\chi^{\{\alpha\}}(R') = f_R'^{(C)}$	(12)

Eq. (8) following immediately from (4) shows that the characters $\chi(R'), \chi^{T}(R'), \chi^{TV}(R'), \chi^{RV}(R')$ must be calculated before $\chi^{V}(R')$. The quantities f_R, $f_R'^{(s)}$, f_R'', $f_R'^{(C)}$ depend only on the rotational part R of R', so that the prime on the subscript can be omitted.

The values of f_R and f_R'' for the various R occurring in space group elements are given in Table 2.

Table 2.

$R \in \mathscr{P}$	Proper rotations, det R = +1					Improper rotations, det R = −1				
	E	C_2	$C_3,\bar{C}_3$	$C_4,\bar{C}_4$	$C_6,\bar{C}_6$	i	σ	$\bar{S}_6,S_6$	$\bar{S}_4,S_4$	$\bar{S}_3,S_3$
f_R'	+3	−1	0	+1	+2	−3	+1	0	−1	−2
f_R''	+6	+2	0	0	+2	+6	+2	0	0	+2

The elements in the right hand part of Table 2 result from the corresponding ones on the left by multiplication with the inversion operator i. $\bar{C}_3 = C_3^2$ is equal to C_3^{-1} etc.

Table 3 gives the values for $f_R''^{(s)}$ and $f_R'^{(C)}$. $f_R''^{(s)}$ can either be $f_R'^{(A)}$ or $f_R'^{(B)}$ or $f_R'^{(C)}$ where A, B, C refer to the classification "point-shaped", "linear", and "nonlinear" given in Sect. 2.3 and are fixed by the first part 1, 2, ... of the double index s of the respective molecule M_s. In the case of linear molecules and R = C_2 or σ the value of $f_R'^{(B)}$ depends on the orientation of the molecular axis ζ with respect to the crystal's C_2-axis (or -axes) and σ-plane(s). In this case the orientation of ζ in space according to Sect. 2.3 will be expressed by an additional second part $a, b, ...$ of the index s.

Table 3.

$R \in \mathscr{P}$		Proper rotations					Improper rotations				
		E	C_2 $\zeta\|C_2$ $\zeta\bot C_2$	$C_3,\bar{C}_3$	$C_4,\bar{C}_4$	$C_6,\bar{C}_6$	i	σ $\zeta\bot\sigma$ $\zeta\|\sigma$	$\bar{S}_6,S_6$	$\bar{S}_4,S_4$	$\bar{S}_3,S_3$
point	$f_R'^{(A)}$	0	0 0	0	0	0	0	0 0	0	0	0
linear	$f_R'^{(B)}$	+2	−2 0	−1	0	+1	+2	−2 0	−1	0	+1
nonlinear	$f_R'^{(C)}$	+3	−1	0	+1	+2	+3	−1	0	+1	+2

If a direction group element R specified in Table 3 is not a symmetry operation for the molecule M_s in question, which can happen only for operations $R \in \mathscr{P}$ not contained in the site group $\mathscr{L}$ belonging to the COM position of the molecule (see Sect. 5), $f_R'^{...}$ does not play any role for FGA (in (7) the respective $N_{M_s}^{pc}(R')$ vanishes) and its value can be left open.

The entries in Tables 2 and 3 have been calculated by means of the general formulae (upper/lower sign for proper/improper rotations R; ζ molecular axis; φ_R and $\cos\varphi_R$ given in Table 4):

$$f_R = \pm 1 + 2 \cos \varphi_R \tag{13}$$

$$f_R'^{(A)} = 0 \tag{14}$$

$$f_R'^{(B)} = \begin{cases} 0 \text{ for } R = C_2 \perp \zeta \text{ and } R = \sigma \parallel \zeta \\ \pm 2 \cos \varphi_R \text{ for all other symmetry operations of} \\ \qquad\qquad\qquad \text{the linear molecule} \end{cases} \tag{15}$$

$$f_R'^{(C)} = +1 \pm 2 \cos \varphi_R = \pm f_R \tag{16}$$

$$f_R'' = 2 \cos \varphi_R (\pm 1 + 2 \cos \varphi_R) = 2 f_R \cos \varphi_R \tag{17}$$

Table 4.

$R \in \mathscr{P}$	E	C_2	$C_3, \bar{C}_3$	$C_4, \bar{C}_4$	$C_6, \bar{C}_6$
	σ	i	$S_3, \bar{S}_3$	$S_4, \bar{S}_4$	$S_6, \bar{S}_6$
φ_R	0	π	$\dfrac{2\pi}{3}$	$\dfrac{\pi}{2}$	$\dfrac{\pi}{3}$
$\cos \varphi_R$	$+1$	-1	$-\dfrac{1}{2}$	0	$+\dfrac{1}{2}$

The elements in the second row result from those in the first row by multiplication with $\sigma = \sigma_h$.

5. Execution of FGA, Inspection Method

FGA is performed by inserting the characters $\chi'''(R')$ of Table 1 in the well-known ("magic") counting formula

$$n_p''' = \frac{1}{[\mathscr{P}']} \sum_{R' \in \mathscr{P}'} \chi'''(R') \chi^{(p)*}(R') = \frac{1}{[\mathscr{P}']} \sum_{q=1}^{r} k_q \chi_q''' \chi_q^{(p)*}. \tag{18}$$

Here $[\mathscr{P}'] = [\mathscr{P}]$ is the order of the factor (or of the direction) group. $\chi^{(p)}(R')$ is the character of the element $R' \in \mathscr{P}'$ (or, by virtue of the isomorphism $\mathscr{P}' \leftrightarrow \mathscr{P}$, of the corresponding element $R \in \mathscr{P}$) in the irrep $\mathcal{D}^{(p)}$ of the point group $\mathscr{P}' \leftrightarrow \mathscr{P}$; it can easily be gathered from the customary point group character tables. q designates that class of conjugate elements in $\mathscr{P}'$ (or $\mathscr{P}$) which contains R' (or R respectively). k_q is the number of elements in this class. The right hand part of (18) presupposes the validity of

$$\chi'''(R') = \chi_q''', \qquad \chi^{(p)}(R') = \chi_q^{(p)} \quad \text{for all } R' \in \text{class } q.$$

r is the total number of classes in $\mathscr{P}'$ (or $\mathscr{P}$). * denotes complex conjugation. The real integer $n_p''' \geq 0$ expresses how often $\mathcal{D}^{(p)}$ shows up in the reduction of $\mathcal{D}'''$. The goal of FGA is the determination of these numbers n_p'''

for all irreps $\mathcal{D}^{(p)}$ of $\mathscr{P} \leftrightarrow \mathscr{P}$ and all reps $\mathcal{D}^{...}$ listed in Table 1.

The greatest difficulties in the practical execution of this program are encountered in determining the particle numbers $N_{M_s}^{PC}(R')$ and $N_A^{PC}(R')$ which are needed in Table 1. It is emphasized that they can always be found by the safe, but often rather unwieldy and time-consuming method of directly inspecting the crystal structure and studying the behavior of the individual particles under all space group operations $\{R | \tau_R\}$. When using this method, the tools developed in the preceding sections suffice. It is helpful then to notice that the above numbers all vanish if R' (apart from the included indefinite primitive translation) represents a s c r e w r o t a - t i o n or a g l i d e r e f l e c t i o n. The reason is that by these typical space group operations every point of the crystal space is displaced to a new position which never differs from the old one by a primitive transla- tion (inclusive 0). The above particle numbers can be different from zero only if R' (apart from a primitive translation) is a point group operation. It is this observation which led to the recognition of a procedure which in many cases considerably facilitates the execution of FGA. This procedure uses the concept of site groups. It will be discussed in Sect. 7.

6. Special Cases

The validity of the character formulae entered in Table 1 and presented without derivation can be tested by considering a variety of special cases, e. g. the following:

a) The values $\chi^{...}(E')$ are equal to the number of degrees of freedom of the respective types of crystal motion. For calculation use column E in Tables 2 and 3 and the remark about case $R' = E'$ in Sect. 2. 4.

b) The case $N_M^{PC} = 1$ (one molecule in the PC) corresponds to the case of one free molecule (forget all PCs except one). The FGA described above then goes over into the well-known method of classifying the normal vibrations of one molecule by their irreps (point group analysis). In this case $\mathscr{P}'$ is isomorphic with the molecular point group $\mathscr{M}$; $R' \in \mathscr{P}'$ can be replaced by $R \in \mathscr{M}$; $N_M^{PC}(R') = 1, \chi^{TV}(R') = 0, N_A^{PC}(R') = N_A^M(R')$ for all $R' \in \mathscr{P}'$; the superscript RV (rotatory vibrations) can be replaced by R (rotations).

c) If $N_A^{M_s} = 1$ for all molecules M_s, the subscripts M_s and M can be replaced by A. Then $\chi^T(R') + \chi^{TV}(R') = \chi(R')$ and by $f_R'^{(s)} = f_R'^{(A)} = 0$ (Table 3, all atoms are point-shaped) $\chi^{RV}(R') = 0$ for all R'. Conse- quently also $\chi^V(R') = 0$. There are only translations and translatory vi- brations in this case.

7. FGA, Advanced Methods

7.1. Generalities

For every point P (whether occupied by a particle or not) within the crystal space there exists a site group $\mathcal{L}_P$ belonging to P. $\mathcal{L}_P$ is defined as the set of all symmetry operations (I) leaving P invariant and (II) bringing the crystal as a whole into coincidence with itself. In brief, $\mathcal{L}_P$ describes the point symmetry of the crystal as seen from point P. By condition (I) $\mathcal{L}_P$ is a point group (for at least P remains fixed under all operations $\in \mathcal{L}_P$) and by condition (II) $\mathcal{L}_P$ is a subgroup of $\mathcal{S}$. The elements of $\mathcal{L}_P$ evidently can then be written in the form

$$\{R_{(P)}|0\} = \{E|r_P\}\{R|0\}\{E|r_P\}^{-1}$$

where r_P is the position vector of P and the point group operation $R_{(P)}$ (rotation part of $\{R_{(P)}|0\}$) is identical with R except having P as a fixed point instead of the origin O. (The symmetry element, e.g. axis or plane, belonging to $R_{(P)}$ is parallel to that of R but goes through P instead of O.) Now all R are contained in the direction group $\mathcal{P}$. Consequently $\mathcal{L}_P$ is isomorphic with a (proper or improper) subgroup of $\mathcal{P}$, in symbols $\mathcal{L}_P \leq \mathcal{P}$, and its order $[\mathcal{L}_P]$ is a divisor of $[\mathcal{P}]$. It is customary to call P a general site if $\mathcal{L}_P = \mathcal{C}_1$ and a special site if $\mathcal{L}_P > \mathcal{C}_1$.

Now imagine all <u>space</u> group operations $\in \mathcal{S}$ being applied to the crystal space. How many <u>different</u> points P_ν, <u>belonging to the same PC</u> as point P will be generated by this procedure from P ? The answer is: Exactly

$$\lambda_P \equiv [\mathcal{P}]/[\mathcal{L}_P], \tag{19}$$

$P \equiv P_1$ inclusive. We shall call these (equivalent) points P_ν, $\nu = 1, ..., \lambda_P$ lying in one PC the (point) grouping Λ_P belonging to P. In this terminology P appears as a representative of all points $P_\nu \in \Lambda_P$. But instead of P any other point P_ν could have been taken as a representative, so that we may write $\Lambda_P \equiv \Lambda_{P_\nu}$, $\nu = 1, ..., \lambda_P$ or, if no confusion is possible, we may omit the subscript P or P_ν on Λ and λ altogether. This idea of point groupings is very important for the understanding of crystal structures for the following reason: Due to the symmetry of the ideal crystal the points P_ν of a grouping Λ either are <u>all</u> unoccupied or <u>all</u> occupied by the CsOM of <u>identical</u> particles (e.g. atoms or molecules of the same chemical and physical species). Furthermore, as the complete crystal with respect to P shows the point symmetry of $\mathcal{L}_P$ the isolated particle situated at P must also show this symmetry (or a higher one), from which it follows that $\mathcal{L}_P$ is a (proper or improper) subgroup of the particle's point group $\mathcal{M}$, in symbols $\mathcal{L}_P \leq \mathcal{M}$.

For later discussions it will be helpful to consider the generation of the points $P_\nu \in \Lambda_P$ from P by application of all operations $\{R|t\} = \{E|a_\alpha\}\{R|\tau_R\} \in \mathscr{S}$ in greater detail. All $\{R|t\}$ with R fixed generate from P a lattice of homologous points, one of which belongs to the same PC as P. Let us call this point $P(R)$. $P(R)$ is one of the points P_ν. In order to associate only $P(R)$ with P and to leave out of consideration the points homologous to $P(R)$ it will be convenient to use as generating operations the factor group elements $R' \equiv \{R|\tau_R\}\mathscr{T} \in \mathscr{P}'$ instead of the above space group elements $\{R|t\} \in \mathscr{S}$, because there is a one-to-one correspondence (denoted by $\sim$) between R' and R and $P(R)$, in symbols $R' \sim R \sim P(R)$. In the special case $R \in \mathscr{L}_P$, $P(R)$ will coincide with P. Consequently, all R contained in the same left coset $R_\nu \mathscr{L}_P$ (R_ν is any representative) of $\mathscr{L}_P$ in $\mathscr{P}$ will be associated with the same point P_ν generated by $R' \sim R$ from P. This confirms again Eq. (19), for $[\mathscr{P}]/[\mathscr{L}_P]$ is the i n d e x (number of cosets) of $\mathscr{L}_P$ in $\mathscr{P}$. Summarizing, P_ν is co-incident with all $P(R)$, $R \in R_\nu \mathscr{L}_P$.

What happens to the absolute orientation (abbreviated $\mathcal{O}$) of a molecule (particularly with the absolute axis ζ orientation of a linear molecule) with COM at P when the operations $R' \in \mathscr{P}'$ are similarly applied? The answer is: Because $\mathscr{L}_P \leq \mathcal{M}$, it does not change (in the sense that it would become distinguishable from the old one) when R is a member of $\mathscr{L}_P$, however it can change (but need not in every case) when $R \notin \mathscr{L}_P$. Applying the above argument about the cosets of $\mathscr{L}_P$ in $\mathscr{P}$ also to this question, we arrive at the result that all R' associated with the elements R of a fixed coset $R_\nu \mathscr{L}_P$ generate the same orientation $\mathcal{O}_\nu$ from $\mathcal{O}$ (this corresponds to the intuitively evident fact that the molecule situated at P_ν is definitely oriented), but that these $\mathcal{O}_\nu$ must not all be different like the corresponding points P_ν. (Indeed in general they will not.)

Now let us consider the site groups $\mathscr{L}_{P_\nu}$ of all the points P_ν of grouping Λ_P. The site group $\mathscr{L}_{P_\nu}$ obviously, due to ideal crystal symmetry, must have the same internal structure as $\mathscr{L}_P$, in other words, it is isomorphic with $\mathscr{L}_P$ in an abstract sense, i. e. when considered as orientationally de-tached from absolute crystal space. However in a more realistic sense, i. e. when regarded in connection with crystal space, the symmetry elements (axes, planes) associated with $\mathscr{L}_{P_\nu}$ may be rotated with respect to those of $\mathscr{L}_P$. Therefore, although $\mathscr{L}_P$ as well as all $\mathscr{L}_{P_\nu}$, $\nu = 1, ..., \lambda_P$ are subgroups of $\mathscr{P}$, they may eventually prove to be non-identical subgroups of $\mathscr{P}$. The elements R_{P_ν} of $\mathscr{L}_{P_\nu}$ result from the elements R_P of $\mathscr{L}_P$ by transformation with the elements $R \in R_\nu \mathscr{L}_P$, in symbols [7]

$$R_{P_\nu} = R R_P R^{-1}, \quad R \in R_\nu \mathscr{L}_P. \tag{20}$$

Proof. Let x be a vector attached to P. Transform it into a corresponding vector x_ν attached to P_ν by

$$x_\nu = Rx, \quad R \in R_\nu \mathcal{L}_P. \tag{21}$$

Assume (20) to be valid. Then, if x by application of R_P goes into

$$\bar{x} \equiv R_P x, \tag{22}$$

x_ν correspondingly by R_{P_ν} goes into

$$\bar{x}_\nu \equiv R_{P_\nu} x_\nu = RR_P R^{-1} Rx = RR_P x = R\bar{x}, \tag{23}$$

which agrees with (21).

By (20)

$$\mathcal{L}_{P_\nu} = R \mathcal{L}_P R^{-1}, \quad R \in R_\nu \mathcal{L}_P, \tag{24}$$

so that $\mathcal{L}_{P_\nu}$ and $\mathcal{L}_P$ are <u>conjugate</u> subgroups in $\mathcal{P}$. If they contain the same elements they are self-conjugate (invariant) in $\mathcal{P}$, or, synonymously, $\mathcal{L}_{P_\nu} = \mathcal{L}_P$ then is a normal divisor of $\mathcal{P}$.

These results apply to <u>all</u> site groups $\mathcal{L}_{P_\nu}$, $\nu = 1,...,\lambda_P$ generated from $\mathcal{L}_P$ by all $R' \in \mathcal{P}'$. All these $\mathcal{L}_{P_\nu}$ are conjugate (perhaps self-conjugate) in $\mathcal{P}$. Their orders all agree: $[\mathcal{L}_{P_\nu}] = [\mathcal{L}_P]$. Regarded as non-ordered sets of point group elements, they are either all identical (case of self-conjugation) or at least partly different. They <u>must</u> be identical when $\mathcal{P}$ because of its abstract structure cannot contain different subgroups of order $[\mathcal{L}_P]$; this can easily be detected by consulting detailed subgroup tables for the point groups. If the $\mathcal{L}_{P_\nu}$ are not all identical as non-ordered sets of elements, let us distinguish the different types occurring in them by the superscript (μ) introducing the following mode of speech: The grouping $\Lambda \equiv \Lambda_P$ is associated with the different c o n c r e t e s i t e g r o u p s $\mathcal{L}^{(\mu)}$, $\mu = 1,...,l$. If $\mathcal{L}_P$ is known, these $\mathcal{L}^{(\mu)}$ can easily be found again by means of a detailed subgroup table for $\mathcal{P}$. Obviously the λ_P i n d i v i d u a l site groups $\mathcal{L}_{P_\nu}$ are divided in equal numbers among the types $\mathcal{L}^{(\mu)}$ so that $l \equiv l_P$ must be a divisor of λ_P and every $\mathcal{L}^{(\mu)}$ is the concrete site group for λ_P / l_P points $P_\nu \in \Lambda_P$. In the limiting case $l_P = 1$ all $\mathcal{L}_{P_\nu}$ are identical (i.e. self-conjugate).

---

[7] Do not confuse R_P with $R_{(P)}$ used in the first paragraph of Sect. 7.1. R_P is simply an abbreviation for any element of the site group $\mathcal{L}_P$ (the subscript P on R_P only identifies the reference point of $\mathcal{L}_P$) while $R_{(P)}$ (or $\{R_{(P)}|0\}$) has the same meaning as R (or $\{R|0\}$) except that point P is fixed and O necessarily is not. In the symbol R_P (and likewise in R_{P_ν} the fact that point P (or P_ν respectively) is fixed instead of O is disregarded. For Eqs. (20) - (24) only the <u>orientation</u> of the symmetry elements associated with R_P and R_{P_ν} plays a role; the parallel shifting in space of the fixed points here is without importance.

The orientation $\mathcal{O}_\nu$ of the molecule situated at P_ν is firmly connected with $\mathcal{L}_{P_\nu}$. The orientation $\mathcal{O}$ of the molecule at P is fixed relative to the symmetry elements associated with $\mathcal{L}_P$. By application of $R' = \{R|\tau_R\}\mathcal{T}$ with $R \in R_\nu\mathcal{L}_P$ to the crystal space, P is changed into P_ν and $R_P \in \mathcal{L}_P$ into $R_{P_\nu} = RR_PR^{-1} \in \mathcal{L}_{P_\nu}$. This implies that the orientation $\mathcal{O}_\nu$ of the molecule at P_ν relative to the symmetry elements of $\mathcal{L}_{P_\nu}$ is the same as $\mathcal{O}$ relative to the symmetry elements of $\mathcal{L}_P$ corresponding by (20) to those of $\mathcal{L}_{P_\nu}$. (The proof is implicit in the equations $\bar{x} \equiv R_P x$ (22) and $\bar{x}_\nu \equiv R_{P_\nu} x_\nu$ (23) above.) Figuratively, $\mathcal{O}$ will rotate with the symmetry elements of the site group when going from P to another point of the grouping Λ_P.

We can now apply these results to FGA. The basic formulae have been summarized in Table 1. In these formulae two partial problems are involved which must be carefully separated: first the determination of the numbers $N_{M_s}^{PC}(R')$ and $N_A^{PC}(R')$ with neglect of molecular orientation (which in certain cases is implicit in the subscript s), and second the determination of these numbers with regard to these orientations. The second problem (concerning only the numbers $N_{M_s}^{PC}(R')$) is more complicated than the first one. Fortunately its solution is only required in the case of linear molecules when simultaneously $\mathcal{P}$ is non-abelian (see below Sect. 7.3) and contains elements C_2, σ and when in addition information about RV and V is wanted. In all other cases (i. e. in the vast majority of all practically occurring cases) the solution of the first problem suffices.

In order to avoid misunderstandings we emphasize again that for the unsophisticated, but tedious inspection method these considerations are unnecessary. What follows is directed exclusively towards the description of a systematic and, in most cases, much more efficient method of FGA.

7.2. First Partial Problem

We shall first discuss the determination of $N_{M_s}^{PC}(R')$ and $N_A^{PC}(R')$ with neglect of molecular orientations. Under this condition only the knowledge of the concrete site groups $\mathcal{L}^{(\mu)}$ belonging to the groupings Λ is needed. As we have seen, the λ_P points of the grouping Λ_P divide in equal numbers λ_P/l_P among the $\mathcal{L}^{(\mu)}$, $\mu = 1, ..., l_P$. We therefore need only distribute the atoms and molecules in the PC among their site groupings Λ and, within each Λ, among the $\mathcal{L}^{(\mu)}$ associated with Λ.

In practice the following procedure is recommended.

Step 1   Let a certain crystal be given. First look for its space group $\mathcal{S}$ and the number of chemical formula units in the PC.

For many crystals these and other data are collected in Wyckoff

[4] and Landolt-Börnstein (abbreviated LB) [5].

Step 2   Then decide which atoms may be considered as forming "mole-
         cules".

         As has already been mentioned in Sect. 2.3 the gathering of the
         PC atoms to larger complexes (molecules) cannot always be done
         without arbitrariness because the interatomic forces in the crystal
         frequently are similar in strength and often can be estimated only
         in a very rough manner. (In a positive sense this arbitrariness re-
         flects the flexibility of FGA.) Tentatively, at least those com-
         plexes which are known to be stable in the free state (e.g. in
         gases or solutions) can be assumed to survive also as essentially
         independent units in the crystal. A physical criterion for the va-
         lidity of this assumption in individual cases is the approximate
         agreement of the IR and R frequencies observed for the vibrations
         of the free molecule and for the corresponding internal vibrations
         of the crystal. "Corresponding" means in this connection that
         there must be a correlation of the respective vibrational irreps of
         $\mathcal{M}$ (point group of free molecule), $\mathcal{L}$ (site group belonging to
         the COM position of the molecule in the crystal), and $\mathcal{P}$ (direc-
         tion group) according to the relation $\mathcal{M} \geq \mathcal{L} \leq \mathcal{P}$.

Step 3   Next determine the groupings Λ for the COM positions of the
         atoms and molecules in the PC.

         It is helpful for this purpose to consult the International Tables
         (abbreviated IT) [6] and Wyckoff [4]. In the IT, p. 73 - 346
         (in the first three columns) the "number of positions, Wyckoff
         notation, and point symmetry" are given for every space group.
         These data refer to our definitions as follows: In the "Wyckoff
         notation" the different types of point groupings Λ possible for the
         respective space group are designated. (Two point groupings Λ_P
         and $\Lambda_{P'}$ are considered as belonging to different types if they can-
         not be transferred into one another by continuously shifting P to-
         wards P' without abruptly changing the site symmetry.) The
         "number of positions" is equal λ multiplied by the ratio of cry-
         stallographic cell volume to PC volume. The "point symmetry" is
         the abstract point group $\mathcal{L} \equiv \mathcal{L}_{P_\nu}$, $P_\nu \in \Lambda$ (in international no-
         tation). It should be noted that the distinction of the "concrete"
         $\mathcal{L}^{(\mu)}$ we need is not made in the IT. The Wyckoff grouping types
         Λ can often be read immediately from Wyckoff [4]. If not, a
         more detailed study of the crystal structure, advantageously using
         the IT, is necessary. Hints are given by the concept of economi-
         cal space filling by the atoms, by chemical bonding properties,

and by the theorem that there must be at least as many atoms or molecules of the respective species in the PC as there are points in the grouping conjectured.

Step 4   The next step is the determination of the concrete conjugate site groups $\mathcal{L}^{(\mu)}$ associated with each Λ in question. This is done by first looking for the <u>abstract</u> point group isomorphic with all these $\mathcal{L}^{(\mu)}$ (it is listed in the IT along with the Wyckoff notation for the type of Λ), then determining a single individual $\mathcal{L}_P$ for one site P and finally deriving from $\mathcal{L}_P$ all point groups conjugate to it in $\mathcal{P}$ (eventually using subgroup tables for the point groups).

Step 5   Then write down the group elements R of <u>every $\mathcal{L}^{(\mu)}$ separately</u>. (If R occurs in several $\mathcal{L}^{(\mu)}$, it must be written repeatedly.) Attribute to every R noted in this list the number of λ_P/l_P particles of the species occupying Λ_P.

Step 6   Finally $N_A^{PC}(R')$ is the sum of all these numbers λ/l noted for <u>any</u> atom A and for R appearing in $R' = \{R|\boldsymbol{\tau}_R\}\mathcal{J}$. Likewise $N_{M_s}^{PC}(R')$ is the sum of all these numbers λ/l noted for the molecules of species M_s and for R corresponding to R'.

It can be seen immediately that the above procedure for determining the numbers $N_A^{PC}(R')$ and $N_{M_s}^{PC}(R')$ differs for individual crystals only in so far as the theoretically possible Wyckoff sites (or grouping types Λ) are differently occupied by atoms or molecules. All the rest of the work can be anticipated and put into tabular form once forever. If a grouping of any Wyckoff type Λ in any $\mathcal{S}$ is assumed to be occupied by a total of λ abstract particles X (one per point $P_\nu \in \Lambda$), the numbers $N_\Lambda^{PC}(R')$ of particles X invariant under operation R' can be calculated at once. This can be done in advance for all Λ and all $\mathcal{S}$. For applications these numbers only must be multiplied by factors $g_{A_{s'}}^\Lambda$,

$$g_A^\Lambda = \sum_{s'} g_{A_{s'}}^\Lambda, \tag{25}$$

or $g_{M_s}^\Lambda$ which indicate how often in the real case the grouping type Λ is occupied by atoms $A_{s'}$ of species s', by atoms A generally, or by molecules M_s of species s, and then added over Λ to give

$$N_{A_{s'}}^{PC}(R') = \sum_\Lambda g_{A_{s'}}^\Lambda N_\Lambda^{PC}(R'), \tag{26}$$

$$N_A^{PC}(R') = \sum_\Lambda g_A^\Lambda N_\Lambda^{PC}(R'), \tag{27}$$

$$N_{M_s}^{PC}(R') = \sum_\Lambda g_{M_s}^\Lambda N_\Lambda^{PC}(R'). \tag{28}$$

The tabulation work can even be taken one step farther. In addition to

$$n_p^T = n_p^\mu = \frac{1}{[\mathcal{P}']} \sum_{R' \in \mathcal{P}'} f_R \chi^{(p)*}(R') = \frac{1}{[\mathcal{P}']} \sum_q k_q f_R \chi_q^{(p)*}, \tag{29}$$

$$n_p^{[\alpha]} = \frac{1}{[\mathcal{P}']} \sum_{R' \in \mathcal{P}'} f_R'' \chi^{(p)*}(R') = \frac{1}{[\mathcal{P}']} \sum_q k_q f_R'' \chi_q^{(p)*}, \tag{30}$$

$$n_p^{\{\alpha\}} = \frac{1}{[\mathcal{P}']} \sum_{R' \in \mathcal{P}'} f_R'^{(C)} \chi^{(p)*}(R') = \frac{1}{[\mathcal{P}']} \sum_q k_q f_R'^{(C)} \chi_q^{(p)*} \tag{31}$$

obviously the following quantities can be calculated in advance for all $\mathcal{S}$ and Λ

$$n_{\Lambda,p} = \frac{1}{[\mathcal{P}']} \sum_{R' \in \mathcal{P}'} N_\Lambda^{PC}(R') f_R \chi^{(p)*}(R'), \tag{32}$$

$$n_{\Lambda,p}'^{(s)} = \frac{1}{[\mathcal{P}']} \sum_{R' \in \mathcal{P}'} N_\Lambda^{PC}(R') f_R'^{(s)} \chi^{(p)*}(R'). \tag{33}$$

Eq. (33) should be considered as a multitude of equations distinguished by the first part of the superscript s. The cases where the second part of s would be needed are excepted here in Sect. 7.2. Incidentally in case A

$$n_{\Lambda,p}'^{(s)} = 0 \quad \text{for all } \Lambda \text{ and } p. \tag{34}$$

Now by insertion of formulae (6) to (9) in (18) it is seen after interchanging the summations occurring in (6), (7), (8) with the summation over R' or q in (18) and after applying (27), (28) that the following relations hold

$$n_p^{TV} = \sum_{s,\Lambda} g_{M_s}^\Lambda n_{\Lambda,p} - n_p^T = \sum_\Lambda g_M^\Lambda n_{\Lambda,p} - n_p^T, \tag{35}$$

$$n_p^{RV} = \sum_{s,\Lambda} g_{M_s}^\Lambda n_{\Lambda,p}'^{(s)}, \tag{36}$$

$$n_p = \sum_\Lambda g_A^\Lambda n_{\Lambda,p}, \tag{37}$$

$$n_p^V = n_p - n_p^T - n_p^{TV} - n_p^{RV}. \tag{38}$$

In (35)

$$g_M^\Lambda = \sum_s g_{M_s}^\Lambda \tag{39}$$

has been used.

The anticipatory tabulation work mentioned above has already been performed by Adams and Newton [7]. Their computer calculated tables give the quantities $n_{\Lambda,p}$ and $n_{\Lambda,p}'^{(s)}$ (the latter integers for case C and the above treated cases B; case A is covered by (34)). All that is left to do for routine FGA is the determination of the factors g_A^Λ and $g_{M_s}^\Lambda$ in the individual case and the simple calculation of $n_p^{\cdots}$ by (29) - (31), (35) - (39). In the examples given in Sect. 8 the use of the tables of Adams and Newton will be also demonstrated.

7.3. Second Partial Problem

In the second partial problem the complications arising from the orientation of the axis ζ of the linear molecules with respect to the C_2 axes or σ planes contained in $\mathscr{P}$ must be coped with. In this case it is no longer sufficient to distinguish the site groups $\mathscr{L}_{P_\nu}$, $P_\nu \in \Lambda_P$, only as non-ordered sets of elements by the $\mathscr{L}^{(\mu)}$ classification, because the (absolute) orientation of ζ in crystal space can be different even for those P_ν whose $\mathscr{L}_{P_\nu}$ contain the same elements as one $\mathscr{L}^{(\mu)}$. This can be seen as follows.

Take $\mathscr{L}^{(\mu)}$ as fixed and assume that $\mathscr{L}_{P_\nu}$ and $\mathscr{L}_{P_{\bar\nu}}$, $P_\nu \neq P_{\bar\nu}$, contain the same elements as $\mathscr{L}^{(\mu)}$, so that in the symbolism of group theory

$$\mathscr{L}_{P_\nu} = \mathscr{L}_{P_{\bar\nu}} = \mathscr{L}^{(\mu)}. \tag{40}$$

Let P_ν be generated from P by $R' \sim$ <u>any</u> $R \in R_\nu \mathscr{L}_P$ and similarly $P_{\bar\nu}$ from P by $\bar R' \sim$ <u>any</u> $\bar R \in R_{\bar\nu} \mathscr{L}_P$ ($\sim$ means one-to-one correspondence; R_ν and $R_{\bar\nu}$ are representatives of the left cosets $R_\nu \mathscr{L}_P$ and $R_{\bar\nu} \mathscr{L}_P$ of $\mathscr{L}_P$ in $\mathscr{P}$). It is evident that $P_{\bar\nu} \neq P_\nu$ only if $R_{\bar\nu} \notin R_\nu \mathscr{L}_P$, i. e. if the two cosets are <u>different</u>. Construct the elements $R_{P_\nu} \in \mathscr{L}_{P_\nu}$ and $R_{P_{\bar\nu}} \in \mathscr{L}_{P_{\bar\nu}}$ by transforming $R_P \in \mathscr{L}_P$ with certain definite elements $R \in R_\nu \mathscr{L}_P$ and $\bar R \in R_{\bar\nu} \mathscr{L}_P$ according to (20) so that

$$R_{P_\nu} = R R_P R^{-1}, \quad R_{P_{\bar\nu}} = \bar R R_P \bar R^{-1}.$$

Then

$$R_{P_{\bar\nu}} = \bar R R^{-1} R_{P_\nu} (\bar R R^{-1})^{-1}. \tag{41}$$

(40) and (41) show that R_{P_ν} and $R_{P_{\bar\nu}}$, both being elements of $\mathscr{L}^{(\mu)}$, are conjugate. As the two cosets were assumed to be different, $\bar R R^{-1} \notin \mathscr{L}_P$ and therefore certainly $\neq E$, and $R_{P_\nu}, R_{P_{\bar\nu}}$ in general will be different. If R_{P_ν} runs through all elements of $\mathscr{L}_{P_\nu} = \mathscr{L}^{(\mu)}$, $R_{P_{\bar\nu}}$ will also run through $\mathscr{L}_{P_{\bar\nu}} = \mathscr{L}^{(\mu)}$, but in general in different succession. Thus when going from P_ν to $P_{\bar\nu}$ the elements of $\mathscr{L}^{(\mu)}$ (which is the common site group for P_ν and $P_{\bar\nu}$), if thought of as an <u>ordered</u> set, in general are permuted in some way.

This change of order may destroy the coincidence of the <u>absolute</u> directions of ζ at P_ν and $P_{\bar\nu}$ because only the orientation of ζ <u>relative to</u> conjugate elements (e. g. R_{P_ν} and $R_{P_{\bar\nu}}$) in the site groups ($\mathscr{L}_{P_\nu}$ and $\mathscr{L}_{P_{\bar\nu}}$) of different grouping points P_ν and $P_{\bar\nu}$ is invariant. Therefore, when going from P_ν to $P_{\bar\nu}$, we must not restrict ourselves to the consideration of the groups $\mathscr{L}_{P_\nu}$ and $\mathscr{L}_{P_{\bar\nu}}$ alone which by (40) coincide as non-ordered sets of elements, but we must carefully watch which individual elements $R_{P_{\bar\nu}} \in \mathscr{L}_{P_{\bar\nu}}$ and $R_{P_\nu} \in \mathscr{L}_{P_\nu}$ are mutually conjugate by (41).

In practice the following procedure will be suitable: Write down all elements $R_{P_\nu} \in \mathscr{L}_{P_\nu} = \mathscr{L}^{(\mu)}$ in some order. Then write underneath every R_{P_ν} that element $\bar R_{P_{\bar\nu}}$ which is conjugate to R_{P_ν}. Do that for every $\mathscr{L}_{P_{\bar\nu}}$. As we have λ_P / l_P different points P_ν whose common concrete site group

is $\mathcal{L}^{(\mu)}$, there will be λ_P/l_P rows of $[\mathcal{L}^{(\mu)}] = [\mathcal{P}]/\lambda_P$ elements each in this
e l e m e n t   t a b l e , and in every row the same elements will appear, but
in general in different order. Now instead of cumulatively attributing to
every element of $\mathcal{L}^{(\mu)}$ λ_P/l_P linear molecules occupying the grouping Λ_P
as in Sect. 7.2 (step 5), only one linear molecule should be attributed to
each element in every row of the element table described above. Then
for the elements in the same column of this table the factors $f_R^{\prime(s)}$ needed
in Eq. (7) of Table 1 will have the same value. The proof of the last statement is implicit in the invariance of the ζ direction relative to the local site symmetry elements:

$$f_{R_{P_\nu}}^{\prime\,(s)} = f_{R_{P_{\bar\nu}}}^{\prime\,(\bar s)} , \tag{42}$$

where R_{P_ν} and $R_{P_{\bar\nu}}$ are conjugate elements (see (41)) of $\mathcal{L}_{P_\nu}$ and $\mathcal{L}_{P_{\bar\nu}}$, and
$s, \bar s$ (to be distinguished by their second parts $a, b, \dots$ only) refer to the
molecules at P_ν , $P_{\bar\nu}$, respectively. (42) particularly is valid in the special case of interest here, i. e. if $R_{P_\nu}, R_{P_{\bar\nu}}$ are two-fold rotations or reflections.

If $\mathcal{P}$ is abelian, (41) gives $R_{P_{\bar\nu}} = R_{P_\nu}$, the elements in the columns of
the element table are then not only conjugate but identical and we can
return to the simpler method of Sect. 7.2. Table 5 shows for convenience
which $\mathcal{P}$ occurring in crystallography are abelian and which not.

Table 5.

Crystal system (syngony)	Lattice point group (holohedry) $\mathcal{P}_{max}$	Crystal class Direction group $\mathcal{P} \leq \mathcal{P}_{max}$		Number
		abelian	non-abelian	
Triclinic	$\mathcal{C}_i$	$\mathcal{C}_1$, $\mathcal{C}_i$		2
Monoclinic	$\mathcal{C}_{2h}$	$\mathcal{C}_2, \mathcal{C}_s, \mathcal{C}_{2h}$		3
Orthorhombic	$\mathcal{D}_{2h}$	$\mathcal{C}_{2v}, \mathcal{D}_2, \mathcal{D}_{2h}$		3
Tetragonal	$\mathcal{D}_{4h}$	$\mathcal{C}_4, \mathcal{S}_4, \mathcal{C}_{4h}$	$\mathcal{D}_4, \mathcal{C}_{4v}, \mathcal{D}_{2d}, \mathcal{D}_{4h}$	7
Trigonal (or rhombohedral)	$\mathcal{D}_{3d}$	$\mathcal{C}_3, \mathcal{S}_6$	$\mathcal{D}_3, \mathcal{C}_{3v}, \mathcal{D}_{3d}$	5
Hexagonal	$\mathcal{D}_{6h}$	$\mathcal{C}_6, \mathcal{C}_{3h}, \mathcal{C}_{6h}$	$\mathcal{D}_6, \mathcal{C}_{6v}, \mathcal{D}_{3h}, \mathcal{D}_{6h}$	7
Cubic	$\mathcal{O}_h$		$\mathcal{J}, \mathcal{J}_h, \mathcal{J}_d, \mathcal{O}, \mathcal{O}_h$	5
Total number	7	16	16	32

In the Tables of Adams and Newton [7] supplementary tables are included (called Tables 3 and 4) which in a somewhat indirect way can also
be used to find n_p^{RV} and n_p^V for the cases of linear molecules contained
in the PC (see Examples 8.4 and 8.5).

8. Examples

The following examples do not contain any new results. They have been
chosen deliberately such that comparison with the methods applied by
other authors is possible. The examples are presented only to illustrate
the concepts developed and the procedure described in Sections 5 and 7.
In order to facilitate comparison with the abstract rules given the details
are given far more circumstantially than actually needed in routine work.
The examples should all be studied because each one presents other views
of the outlined theory.

In Tables 6, 10, 14, 17, 22 the columns are numbered 1 to 18. Columns 1 to 9 contain the informations drawn from crystallography for the
respective example, columns 10 to 13 refer to the atoms and 14 to 18 to
the molecules present in a PC. In the individual columns the following
data are collected:

(1) Λ type of grouping in Wyckoff notation

(2) $\mathcal{L}$ abstract site group in Schoenflies and international notation

(3) $[\mathcal{L}]$ site group order

(4) $\lambda = [\mathcal{P}]/[\mathcal{L}]$ number of points P_ν per grouping Λ

(5) l number of concrete site groups $\mathcal{L}^{(\mu)}$

(6) λ/l number of grouping points P_ν with the same concrete
           site group $\mathcal{L}^{(\mu)}$

(7) $\mathcal{L}^{(\mu)}; \mu = 1,\dots,l$ concrete site group

(8) $\mathcal{L}_{P_\nu}; \nu = 1,\dots,\lambda$ individual site group (column needed only for
           linear molecules)

(9) $R \in \mathcal{L}^{(\mu)}$ elements of concrete site group $\mathcal{L}^{(\mu)}$ or, if needed,
     $R \in \mathcal{L}_{P_\nu}$ elements of individual site group $\mathcal{L}_{P_\nu}$ (conjugate elements
           below one another)

(10) A_s chemical symbol for the atom

(11) $g^\Lambda_{A_s}$ number of site groupings of type Λ occupied by atoms A_s

(12) $\frac{\lambda}{l} g^\Lambda_{A_s}$ number of atoms A_s invariant under the operations $R' \sim$
           $\sim R \in \mathcal{L}^{(\mu)}$

(13) $\lambda g^\Lambda_A = \lambda \sum_s g^\Lambda_{A_s}$ total number of atoms A in the site groupings of
           type Λ

(14) M_s abbreviated symbol (see text) for the molecule chosen

(15) $g_{M_s}^\Lambda$ number of site groupings of type Λ occupied by molecules M_s [8]

(16) $\frac{\lambda}{l} g_{M_s}^\Lambda$ number of molecules M_s with COM invariant under the operations $R' \sim R \in \mathcal{L}^{(\mu)}$ (to be filled when only the concrete site groups in column 7 are needed)

(17) $\frac{\lambda}{l} g_{M_s}^\Lambda$ number of molecules M_s with COM invariant under the operations $R' \sim R \in \mathcal{L}_{P_\nu}$ (to be filled instead of column 16 when the individual site groups in column 8 are needed)

(18) $\lambda g_M^\Lambda = \lambda \sum_s g_{M_s}^\Lambda$ total number of molecules M in the site groupings of type Λ

The numbers which are needed for insertion in Tables 7, 11, 15, 18, 23 are encircled.

8.1. Example 1: Tolane

(Wyckoff Vol. 6/2 p. 94; Schrader [2] p. 12; LB Vol. I/4 p. 222, 392)

$S = C_{2h}^5 = P2_1/a$ (space group No. 14), $\mathcal{P} = C_{2h} = 2/m$ (abelian), $[\mathcal{P}] = 4$.

The PC is identical with the monoclinic crystallographic cell Γ_m and contains 4 formula units $C_{14}H_{10}$ (or $C_6H_5C \equiv CC_6H_5$) which will be considered as the only "molecules" $M_1 = M$ in the PC. The Wyckoff sites for the molecular CsOM can be taken arbitrarily as a and b or c and d. In Table 6 the occupation of the groupings Λ by the atoms A (= carbon C or hydrogen H) and molecules M (= $C_{14}H_{10}$) of a PC is given in detail. Fig. 1 shows the PC and one grouping each of a, b, e.

Table 6 immediately yields the numbers $N_A^{PC}(R')$ and $N_M^{PC}(R')$ needed in Table 1. The site group elements $R \in \mathcal{L}^{(\mu)}$ in column 9, if primed, are those elements $R' \in \mathcal{P}'$ for which alone these numbers are $\neq 0$. Moreover, the <u>value</u> of these numbers is then simply the sum of the numbers of atoms or molecules given in columns 12 and 16. Thus

$N_A^{PC}(E') = 96, \quad N_A^{PC}(R' \neq E') = 0;$

$N_M^{PC}(E') = 4, \quad N_M^{PC}(i') = 4, \quad N_M^{PC}(R' \neq E', i') = 0.$

---

[8] Every grouping must be filled completely by molecules of the same chemical and physical species, but not of equal orientations in space. Therefore fractional numbers $g_{M_s}^\Lambda$ are allowed when s is specified not only by $1, 2, \ldots$ (first part) but also by $a, b, \ldots$ (second part).

Fig. 1.

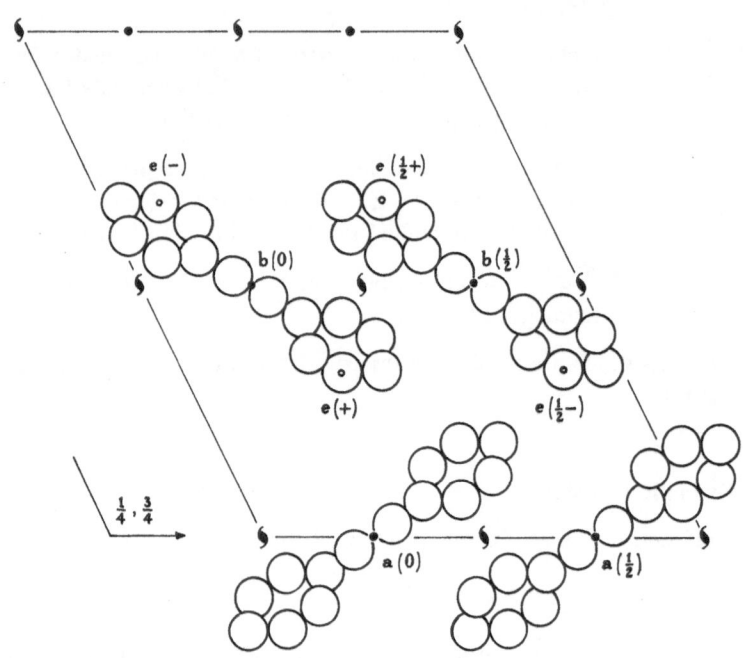

Table 6.

1	2	3	4	5	6	7	8	9	10	11	12	13	14	15	16	17	18
Λ	$\mathcal{L}$	$[\mathcal{L}]$	λ	l	$\frac{\lambda}{l}$	$\mathcal{L}^{(\mu)}$		$R \in \mathcal{L}^{(\mu)}$	A_s	$g^{\Lambda}_{A_s}$	$\frac{\lambda}{l} g^{\Lambda}_{A_s}$	λg^{Λ}_{A}	M_s	$g^{\Lambda}_{M_s}$	$\frac{\lambda}{l} g^{\Lambda}_{M_s}$		λg^{Λ}_{M}
e	$C_1 \equiv 1$	1	4	1	4	C_1		E	C	14	56	96					
									H	10	40						
a	$C_i \equiv \bar{1}$	2	2	1	2	C_i		E, i					M	1	2		2
b	$C_i \equiv \bar{1}$	2	2	1	2	C_i		E, i					M	1	2		2

Using these results and Tables 2 and 3 we now can proceed to fill in the formulae of Table 1 and calculate the irrep numbers $n^{...}_p$ with Eq. (18). The data necessary for the calculation are collected in Table 7. The upper part contains what is to be taken from Table 1 in this case, the lower part contains the values of the irrep characters $\chi^{(P)}(R') = \chi^{(P)}(R)$ to be taken from a character table of point group $\mathcal{P} = C_{2h}$. M being nonlinear, $f^{\prime(s)}_R = f^{\prime(C)}_R$. The result of the calculation with Eq. (18) is given in Table 8.

Table 7.

$\mathcal{P}' = C'_{2h} \equiv C^5_{2h}/\mathcal{T}$ ... $R' =$	E'	C'$_2$	i'	σ'
$N_A^{PC}(R') =$	96	0	0	0
$N_M^{PC}(R') =$	4	0	4	0
$\chi^T(R') = f_R =$	3	-1	-3	1
$\chi^{\{\alpha\}}(R') = f_R^{'(C)} =$	3	-1	3	-1
$\chi^{[\alpha]}(R') = f_R'' =$	6	2	6	2
$\chi(R') = N_A^{PC}(R')f_R =$	288	0	0	0
$\chi^{TV}(R') = (N_M^{PC}(R')-1)f_R =$	9	1	-9	-1
$\chi^{RV}(R') = N_M^{PC}(R')f_R^{'(C)} =$	12	0	12	0
$\chi^V(R') = \chi(R') - \chi^T(R') - \chi^{TV}(R') - \chi^{RV}(R') =$	264	0	0	0
$\chi^{(P)}(R') = \begin{cases} \chi^{A_g}(R') = \\ \chi^{B_g}(R') = \\ \chi^{A_u}(R') = \\ \chi^{B_u}(R') = \end{cases}$	1 1 1 1	1 -1 1 -1	1 1 -1 -1	1 -1 -1 1

Table 8.

p	n_p	$n_p^T = n_p^\mu$	n_p^{TV}	n_p^{RV}	n_p^V	$n_p^{[\alpha]}$	$n_p^{\{\alpha\}}$
A_g	72	0	0	6	66	4	1
B_g	72	0	0	6	66	2	2
A_u	72	1	5	0	66	0	0
B_u	72	2	4	0	66	0	0
Total	288	3	9	12	264	6	3

There are several ways in which the result of the calculations may be checked: n_p must be equal to the sum $n_p^T + n_p^{TV} + n_p^{RV} + n_p^V$ for all p. The totals of the columns $n_p, n_p^T, ...$ after multiplication of the rows with the irrep degeneracy (here everywhere 1) must agree with $\chi(E'), \chi^T(E'), ...$ (see column E' in Table 7). The entries in columns $n_p^T, n_p^{\{\alpha\}}, n_p^{[\alpha]}$ can be compared with the irreps for the translations T_x, T_y, T_z, rotations R_x, R_y, R_z, and polarizability components α_{xx} etc. given in the usual character tables (see e.g. [8], p. 323sq).

With the Tables of Adams and Newton the calculation is done as follows. For space group No. 14 "Table 2" of Adams and Newton lists the data reproduced in Table 9.

Table 9.

14 C_{2h}					Rotatory			
Wyckoff	A_g	B_g	A_u	B_u	A_g	B_g	A_u	B_u
2 a - d	0	0	3	3	3	3	0	0
4 e	3	3	3	3	3	3	3	3

In the first column the numbers 2 and 4 are the values of λ for the Wyckoff sites Λ = a, ..., d and e, respectively. The numbers in the third and fourth row below p = A_g, B_g, A_u, B_u represent the quantities $n_{\Lambda,p}$ (columns 3 through 6) and $n'^{(s)}_{\Lambda,p}$ (columns 7 through 10, below "Rotatory") explained in Sect. 7.2. Having here only one species of molecule we can omit the index s. For calculating n_p, n_p^{TV}, n_p^{RV}, n_p^V by (35) - (39) we need g_A^Λ, $g_{M_s}^\Lambda$ (here $= g_M^\Lambda$), and n_p^T. n_p^T (and likewise $n_p^{[\alpha]}$ and $n_p^{\{\alpha\}}$) must be taken from the customary character tables for point groups, e. g. in Wilson-Decius-Cross [8], p. 326, C_{2h} (for n_p^T use entries T_z in A_u and T_x, T_y in B_u; for $n_p^{[\alpha]}$ use entries α_{xx}, α_{yy}, α_{zz}, α_{xy} in A_g and α_{yz}, α_{zx} in B_g; for $n_p^{\{\alpha\}}$ use entries R_z in A_g and R_x, R_y in B_g; it is seen that the numbers of these entries correspond to Table 8, columns 3, 7, 8). In one PC we have 14 C-atoms and 10 H-atoms on sites e and 1 tolane molecule each on sites a and b; therefore the values of g_A^Λ and g_M^Λ are

$$g_A^e = 14 + 10 = 24, \quad g_M^a = g_M^b = 1,$$ all others vanishing. (These values of course can also be taken from Table 6, columns 11, 15.)

We show the calculation of the numbers $n_p^{..}$ with (35) - (38) schematically:

| | | p | = A_g B_g A_u B_u | | | | | A_g B_g A_u B_u |

(37): n_p (only site e needed) $n_{e,p}$ = 3 3 3 3 times g_A^e = 24 gives 72 72 72 72

(35): $n_p^{TV}+n_p^T$ (only sites a,b needed) $n_{a,p}=n_{b,p}$ = 0 0 3 3 times $g_M^a+g_M^b$ = 2 gives 0 0 6 6

n_p^T (from character tables) 0 0 1 2

n_p^{TV} $= (n_p^{TV}+n_p^T) - n_p^T$ 0 0 5 4

(36): n_p^{RV} (only sites a,b needed; $n'_{a,p}=n'_{b,p}$ = 3 3 0 0 times $g_M^a+g_M^b$ = 2 gives 6 6 0 0 use "Rotatory" in Table 9)

(38): n_p^V $= n_p - (n_p^{TV}+n_p^T) - n_p^{RV}$ 66 66 66 66

$n_p^{[\alpha]}$, $n_p^{\{\alpha\}}$ must be taken from the character tables.

8.2. Example 2: Lithium Iodate

(Wyckoff Vol. 2 p. 387; LB Vol. I/4 p. 91; the space group $\mathfrak{D}_6^6$ given there must be changed into $\mathfrak{C}_6^6$, see [9, 10, 11]).

$\mathscr{S} = \mathfrak{C}_6^6 = P6_3$ (space group No. 173), $\mathscr{P} = C_6 = 6$ (abelian), $[\mathscr{P}] = 6$.

The hexagonal PC Γ_h contains 2 formula units $LiIO_3$. We consider the ions $M_1 = IO_3^-$ (nonlinear) and $M_2 = Li^+$ (point-shaped) as the molecule species filling the PC. Tab. 10 shows the occupation of the Wyckoff sites Λ by the atoms A (= Li, I, O) and the molecular species M_1 and M_2. Fig. 2 shows the crystal structure with the PC and the groupings a, b, c.

Fig. 2.

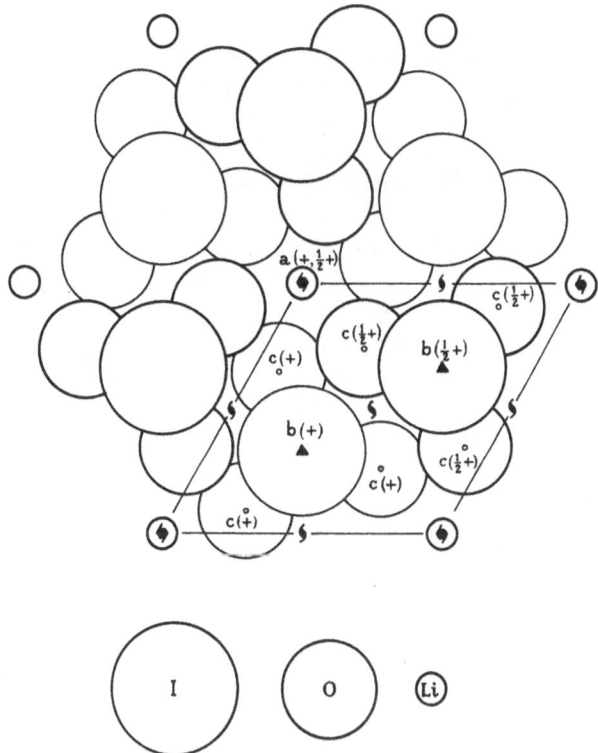

Table 11 collects the data needed for the formulae given in Table 1. As a check on Table 11 it may be noted that by the remark in Sect. 5 all numbers $N_{M_s}^{PC}(R')$ and $N_A^{PC}(R')$ for $R' = C_2', C_6', \bar{C}_6'$ vanish automatically because these factor group elements refer to screw rotations in $\mathfrak{C}_6^6$. The two-dimensional r e a l irreps E_1 and E_2 split into two one-dimensional irreps

Table 10.

1	2	3	4	5	6	7	8	9	10	11	12	13	14	15	16	17	18
Λ	$\mathcal{L}$	$[\mathcal{L}]$	λ	l	$\frac{\lambda}{l}$	$\mathcal{L}^{(\mu)}$		$R\in\mathcal{L}^{(\mu)}$	A_s	$g^\Lambda_{A_s}$	$\frac{\lambda}{l}g^\Lambda_{A_s}$	λg^Λ_A	M_s	$g^\Lambda_{M_s}$	$\frac{\lambda}{l}g^\Lambda_{M_s}$		λg^Λ_M
a	$C_3\equiv 3$	3	2	1	2	C_3		$E,C_3,\bar{C}_3$	Li	1	②	2	M_2	1	②		2
b	$C_3\equiv 3$	3	2	1	2	C_3		$E,C_3,\bar{C}_3$	I	1	②	2	M_1	1	②		2
c	$C_1\equiv 1$	1	6	1	6	C_1		E	O	1	⑥	6					

Table 11.

$\mathscr{P}' = C_6' = C_6^6/\mathscr{T}$ $\qquad$ R' =	E'	C_3'	$\bar{C}_3'$	C_2'	C_6'	$\bar{C}_6'$
$N_A^{PC}(R') =$	10	4	4	0	0	0
$N_{M_1}^{PC}(R') =$	2	2	2	0	0	0
$N_{M_2}^{PC}(R') =$	2	2	2	0	0	0
$N_M^{PC}(R') = \sum_s N_{M_s}^{PC}(R') =$	4	4	4	0	0	0
$\chi^T(R') = f_R =$	3	0	0	-1	2	2
$\chi^{(\alpha)}(R') = f_R'^{(1)} = f_R'^{(C)} =$	3	0	0	-1	2	2
$f_R'^{(2)} = f_R'^{(A)} =$	0	0	0	0	0	0
$\chi^{[\alpha]}(R') = f_R'' =$	6	0	0	2	2	2
$\chi(R') = N_A^{PC}(R')f_R =$	30	0	0	0	0	0
$\chi^{TV}(R') = (N_M^{PC}(R')-1)f_R =$	9	0	0	1	-2	-2
$\chi^{RV}(R') = \sum_s N_{M_s}^{PC}(R')f_R'^{(s)} =$	6	0	0	0	0	0
$\chi^V(R') = \chi(R') - \chi^T(R') - \chi^{TV}(R') - \chi^{RV}(R') =$	12	0	0	0	0	0
$\chi^A(R') =$	1	1	1	1	1	1
$\chi^B(R') =$	1	1	1	-1	-1	-1
$\chi^{E_1^{(1)}}(R') =$	1	$-\varepsilon^*$	$-\varepsilon$	-1	ε	ε^*
$\chi^{E_1^{(2)}}(R') =$	1	$-\varepsilon$	$-\varepsilon^*$	-1	ε^*	ε
$\chi^{E_2^{(1)}}(R') =$	1	$-\varepsilon$	$-\varepsilon^*$	1	$-\varepsilon^*$	$-\varepsilon$
$\chi^{E_2^{(2)}}(R') =$	1	$-\varepsilon^*$	$-\varepsilon$	1	$-\varepsilon$	$-\varepsilon^*$

The lower block of characters is bracketed together as $\chi^{(p)}(R') =$.

each ($E_1^{(1)}$, $E_1^{(2)}$ and $E_2^{(1)}$, $E_2^{(2)}$, respectively) in the complex field. These latter irreps must be used instead of E_1 and E_2 because the theory leading to (18) presupposes the representation space to be complex-valued. ε is an abbreviation for $\exp(\frac{1}{3}\pi i)$.

The irrep numbers resulting from Table 11 and Eq. (18) are shown in Table 12. For all p $d_p = 1$ and consequently $n_p^{\cdots} = d_p n_p^{\cdots}$. ($d_p$ is the dimension of $\mathcal{D}^{(p)}$.)

Table 12.

p	n_p	$n_p^T = n_p^\mu$	n_p^{TV}	n_p^{RV}	n_p^V	$n_p^{[\alpha]}$	$n_p^{\{\alpha\}}$
A	5	1	1	1	2	2	1
B	5	0	2	1	2	0	0
$E_1^{(1)}$	5	1	1	1	2	1	1
$E_1^{(2)}$	5	1	1	1	2	1	1
$E_2^{(1)}$	5	0	2	1	2	1	0
$E_2^{(2)}$	5	0	2	1	2	1	0
Total	30	3	9	6	12	6	3

We now show how to obtain this result by the method of Adams and Newton. Their "Table 2" for space group No. 173 gives the data reproduced in Table 13.

Table 13.

173 $\mathcal{C}_6$					Rotatory			
Wyckoff	A	B	E_1	E_2	A	B	E_1	E_2
2 a - b	1	1	1	1	1	1	1	1
6 c	3	3	3	3	3	3	3	3

The calculation proceeds as follows:

			A	B	E_1	E_2
n_p	2 Li-atoms on Wyckoff site a give	2a	1	1	1	1
	2 I-atoms on Wyckoff site b give	2b	1	1	1	1
	6 O-atoms on Wyckoff site c give	6c	3	3	3	3
		total	5	5	5	5
$n_p^{TV} + n_p^T$	2 molecules M_2 on Wyckoff site a give	2a	1	1	1	1
	2 molecules M_1 on Wyckoff site b give	2b	1	1	1	1
		total	2	2	2	2
n_p^T	(from character tables)		1	0	1	0
n_p^{TV}	$= (n_p^{TV} + n_p^T) - n_p^T$		1	2	1	2
n_p^{RV}	2 polyatomic nonlinear molecules M_1 on Wyckoff site b give 2b (Rot.)		1	1	1	1
	(the point-shaped molecules M_2 here are not counted)					
n_p^V	$= n_p - (n_p^{TV} + n_p^T) - n_p^{RV}$		2	2	2	2
$n_p^{[\alpha]}$	(from character tables, $\alpha_{xx} + \alpha_{yy}$, etc.)		2	0	1	1
$n_p^{\{\alpha\}}$	(from character tables, R_z, etc.)		1	0	1	0

In the last calculation E_1, E_2 are considered to be two-dimensional; therefore the numbers for E_1 and E_2 must be doubled when looking for the total number of the respective degrees of freedom.

8.3. Example 3: Calcite

(Wyckoff Vol. 2 p. 362; LB Vol. I/4 p. 102; Bhagavantam-Venkatarayudu [1] (1969) p. 145 (with other denominations of the irreps); Sushchinsky [2] p. 419 (with some errors in the final result))

$\mathscr{S} = \mathfrak{D}_{3d}^6 = R\bar{3}c$ (space group No. 167), $\mathscr{P} = \mathfrak{D}_{3d} = \bar{3}m$ (non-abelian), $[\mathscr{P}] = 12$.

The trigonal (or rhombohedral) PC Γ_{rh} contains 2 formula units $CaCO_3$. We assume that the two molecular species are $M_1 = CO_3^{--}$ (non-linear) and $M_2 = Ca^{++}$ (point-shaped). The Wyckoff sites occupied by the atoms A (= C, Ca, O) and by these molecules are given in Table 14. Fig. 3 shows schematically the PC and the sites.

Table 14.

1	2	3	4	5	6	7	8	9	10	11	12	13	14	15	16	17	18
Λ	$\mathscr{L}$	$[\mathscr{L}]$	λ	l	$\frac{\lambda}{l}$	$\mathscr{L}^{(\mu)}$		$R \in \mathscr{L}^{(\mu)}$	A_s	$g^\Lambda_{A_s}$	$\frac{\lambda}{l} g^\Lambda_{A_s}$	λg^Λ_A	M_s	$g^\Lambda_{M_s}$	$\frac{\lambda}{l} g^\Lambda_{M_s}$		λg^Λ_M
a	$\mathfrak{D}_3 \equiv 32$	6	2	1	2	$\mathfrak{D}_3$		$E\ C_3\ \bar{C}_3\ C_2^{(1)}C_2^{(2)}C_2^{(3)}$	C	1	(2)	2	M_1	1	2		2
b	$C_{3i} \equiv S_6 \equiv \bar{3}$	6	2	1	2	C_{3i}		$E\ C_3\ \bar{C}_3\ i\ S_6\ \bar{S}_6$	Ca	1	(2)	2	M_2	1	2		2
e	$C_2 \equiv 2$	2	6	3	2	$C_2^{(1)}$ $C_2^{(2)}$ $C_2^{(3)}$		$E\ C_2^{(1)}$; $E\ C_2^{(2)}$; $E\ C_2^{(3)}$	O	1	(2)(2)(2)	6					

Table 15 presents the detailed application of Table 1 on this example ($C_2^{(*)}$ stands for one of $C_2^{(1)}, C_2^{(2)}, C_2^{(3)}$) and Table 16 gives the calculated irrep numbers $n_p^{\cdots}$.

The calculation with the Tables of Adams and Newton does not show anything new in comparison to Example 2 and therefore will be omitted here.

Fig. 3.[9]

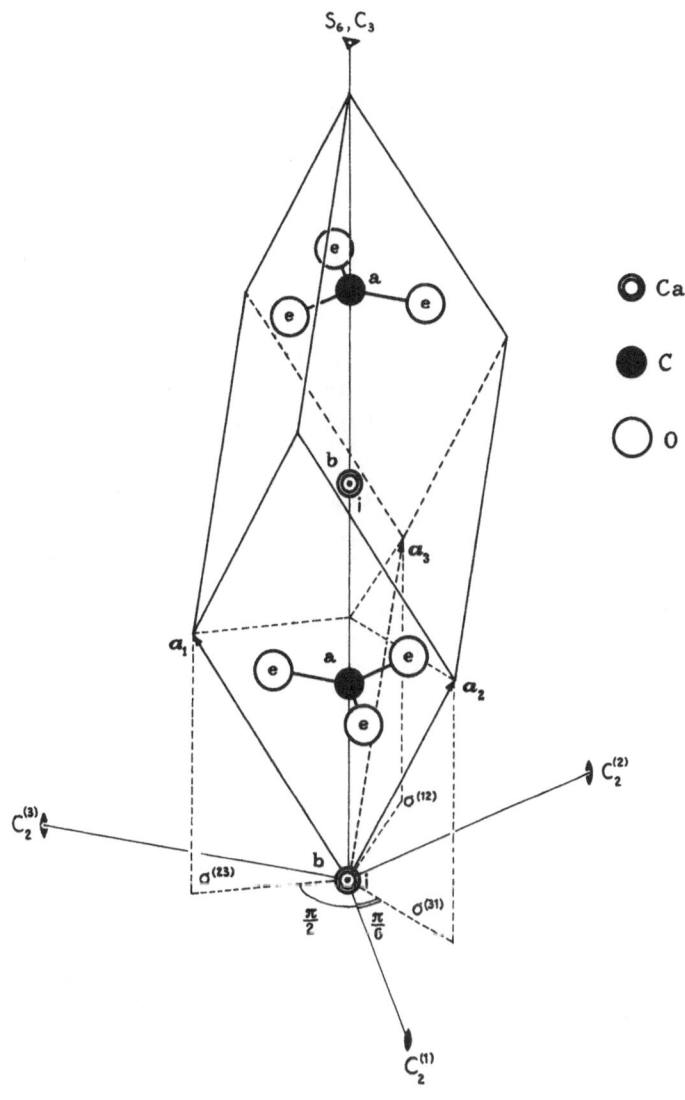

[9] $C_2^{(1)}, C_2^{(2)}, C_2^{(3)}$ only show the <u>directions</u> of the twofold rotation axes which actually go through sites a . $\sigma_d = \sigma^{(23)}, \sigma^{(31)}, \sigma^{(12)}$ are glide planes.

Table 15.

$\mathscr{P}' = \mathscr{D}'_{3d} = \mathscr{D}^6_{3d}/\mathscr{T}$ $\quad k_q R' =$	1E'	2C'$_3$	3C$_2^{(')'}$	1i'	2S'$_6$	3σ'$_d$
$N_A^{PC}(R') =$	10	4	4	2	2	0
$N_{M_1}^{PC}(R') =$	2	2	2	0	0	0
$N_{M_2}^{PC}(R') =$	2	2	0	2	2	0
$N_M^{PC}(R') = \sum_s N_{M_s}^{PC}(R') =$	4	4	2	2	2	0
$\chi^T(R') = f_R =$	3	0	-1	-3	0	1
$\chi^{(\alpha)}(R') = f_R'^{(1)} = f_R'^{(C)} =$	3	0	-1	3	0	-1
$f_R'^{(2)} = f_R'^{(A)} =$	0	0	0	0	0	0
$\chi^{[\alpha]}(R') = f_R'' =$	6	0	2	6	0	2
$\chi(R') = N_A^{PC}(R')f_R =$	30	0	-4	-6	0	0
$\chi^{TV}(R') = (N_M^{PC}(R')-1)f_R =$	9	0	-1	-3	0	-1
$\chi^{RV}(R') = \sum_s N_{M_s}^{PC}(R')f_R'^{(s)} =$	6	0	-2	0	0	0
$\chi^V(R') = \chi(R') - \chi^T(R') - \chi^{TV}(R') - \chi^{RV}(R') =$	12	0	0	0	0	0
$\chi^{(P)}(R') = \begin{cases} \chi^{A_{1g}}(R') = \end{cases}$	1	1	1	1	1	1
$\chi^{A_{2g}}(R') =$	1	1	-1	1	1	-1
$\chi^{E_g}(R') =$	2	-1	0	2	-1	0
$\chi^{A_{1u}}(R') =$	1	1	1	-1	-1	-1
$\chi^{A_{2u}}(R') =$	1	1	-1	-1	-1	1
$\chi^{E_u}(R') =$	2	-1	0	-2	1	0

Table 16.

p	n_p	$n_p^T = n_p^M$	n_p^{TV}	n_p^{RV}	n_p^V	$n_p^{[\alpha]}$	$n_p^{(\alpha)}$	d_p	$d_p n_p$	$d_p n_p^T$	$d_p n_p^{TV}$	$d_p n_p^{RV}$	$d_p n_p^V$	$d_p n_p^{[\alpha]}$	$d_p n_p^{(\alpha)}$
A_{1g}	1	0	0	0	1	2	0	1	1	0	0	0	1	2	0
A_{2g}	3	0	1	1	1	0	1	1	3	0	1	1	1	0	1
E_g	4	0	1	1	2	2	1	2	8	0	2	2	4	4	2
A_{1u}	2	0	1	0	1	0	0	1	2	0	1	0	1	0	0
A_{2u}	4	1	1	1	1	0	0	1	4	1	1	1	1	0	0
E_u	6	1	2	1	2	0	0	2	12	2	4	2	4	0	0
Total								30	3	9	6	12	6	3	

8.4. Example 4: Potassium Azide

(Wyckoff Vol. 2 p. 277; LB Vol. I/4 p. 56; Turrell [2] p. 128)

$\mathcal{S} = \mathcal{D}_{4h}^{18}$ = I4/mcm (space group No. 140), $\mathcal{P} = \mathcal{D}_{4h}$ = 4/mmm (non-abelian), $[\mathcal{P}] = 16$.

The volume of the PC is one half of the body-centered tetragonal crystallographic cell Γ_q^v and contains 2 formula units KN_3. We distinguish the molecule types $M_{1a} = N_3^-$ (linear with axis $\zeta \parallel \hat{C}_2^{(12)}, \sigma^{(12)}, \sigma_h$ and $\perp C_2, C_2^{(21)}, \sigma^{(21)}$), $M_{1b} = N_3^-$ (linear with $\zeta \parallel C_2^{(21)}, \sigma^{(21)}, \sigma_h$ and $\perp C_2, C_2^{(12)}, \sigma^{(12)}$) and $M_2 = K^+$ (point-shaped). Fig. 4 shows the crystal structure and Fig. 5 the elements of $\mathcal{D}_{4h}$. Table 17 gives the sites and Tables 18, 19 summarize the FGA.

Fig. 4.   Fig. 5.

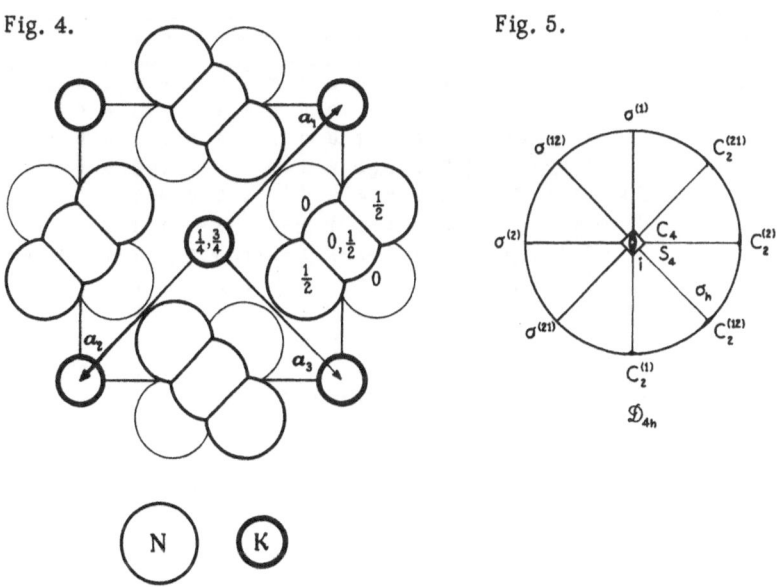

Table 17.

1	2	3	4	5	6	7	8	9	10	11	12	13	14	15	16	17	18
Λ	$\mathcal{L}$	$[\mathcal{L}]$	λ	l	$\frac{\lambda}{l}$	$\mathcal{L}^{(\mu)}$	$\mathcal{L}_{P_\nu}$	$R \in \mathcal{L}_{P_\nu} = \mathcal{L}^{(\mu)}$	A_s	$g^\Lambda_{A_s}$	$\frac{\lambda}{l} g^\Lambda_{A_s}$	λg^Λ_A	M_s	$g^\Lambda_{M_s}$	$\frac{\lambda}{l} g^\Lambda_{M_s}$	$\frac{\lambda}{l} g^\Lambda_{M_s}$	λg^Λ_M
a	$\mathcal{D}_4 \equiv 42$	8	2	1	2	$\mathcal{D}_4$	not needed	$E\ C_4\ C_2\ \bar{C}_4\ C_2^{(1)} C_2^{(12)} C_2^{(12)} C_2^{(21)}$	K	1	②	2	M_2	1	②	not n.	2
d	$\mathcal{D}_{2h} \equiv mmm$	8	2	1	2	$\mathcal{D}_{2h}$	$\mathcal{L}_{P_1} = \mathcal{D}_{2h}$ $\mathcal{L}_{P_2} = C_2 \mathcal{D}_{2h} C_2^1$	$E\ C_2\ i\ \sigma_h\ C_2^{(12)} C_2^{(21)} \sigma^{(12)} \sigma^{(21)}$ $E\ C_2\ i\ \sigma_h\ C_2^{(21)} C_2^{(12)} \sigma^{(21)} \sigma^{(12)}$	N*	1	②	2	M_{1a} M_{1b}	$\frac{1}{2}$ $\frac{1}{2}$	not n.	$\binom{1}{1}^+_+$	2
h	$C_{2v} \equiv mm$	4	4	2	2	$C_{2v}^{(1)}$ $C_{2v}^{(2)}$	not needed "	$E\ C_2^{(12)}\ \sigma_h\ \sigma^{(12)}$ $E\ C_2^{(21)}\ \sigma_h\ \sigma^{(21)}$	N	1	② ②	4					

Footnotes see next page

* Central atom of linear N_3^--molecule.

† The 2×8 elements entered in the rows $\mathscr{L}_{P_1} = \mathfrak{D}_{2h}$ and $\mathscr{L}_{P_2} = C_4 \mathfrak{D}_{2h} C_4^{-1}$ constitute the element table mentioned in Sect. 7.3. Those in the row $\mathscr{L}_{P_2}$ are conjugate to those above them in the row $\mathscr{L}_{P_1}$, e.g. $C_2^{(21)} = C_4 C_2^{(12)} C_4^{-1}$. Conjugation can be effected by $C_4, \bar{C}_4, C_2^{(1)}$ or $C_2^{(2)} \in \mathscr{P}$. The values of $f_R^{\prime(s)}$ for elements standing in the same column of this element table, e.g. $C_2^{(12)}$ for $s = 1a$ and $C_2^{(21)}$ for $s = 1b$ are identical (cf. the entries for $f_R^{\prime(1a)}$ and $f_R^{\prime(1b)}$ in Table 18). To every element of $\mathscr{L}_{P_1}$ and $\mathscr{L}_{P_2}$ 1 molecule M_{1a} or M_{1b}, respectively, (see column 17) must be attributed.

Table 18.[10]

$\mathscr{P}' = \mathfrak{D}'_{4h} = \mathfrak{D}^{18}_{4h}/\mathscr{T}$ R' =	E'	C_4'	$\bar{C}_4'$	C_2'	$C_2^{(1)'}$	$C_2^{(2)'}$	$C_2^{(12)'}$	$C_2^{(21)'}$	I'	S_4'	$\bar{S}_4'$	σ_h'	$\sigma^{(1)'}$	$\sigma^{(2)'}$	$\sigma^{(12)'}$	$\sigma^{(21)'}$
$N_A^{PC}(R') =$	8	2	2	4	2	2	6	6	2	0	0	6	0	0	4	4
$N_{M_{1a}}^{PC}(R') =$	1	0	0	1	0	0	1	1	1	0	0	1	0	0	1	1
$N_{M_{1b}}^{PC}(R') =$	1	0	0	1	0	0	1	1	1	0	0	1	0	0	1	1
$N_{M_2}^{PC}(R') =$	2	2	2	2	2	2	2	2	0	0	0	0	0	0	0	0
$N_M^{PC}(R') = \sum_s N_{M_s}^{PC}(R') =$	4	2	2	4	2	2	4	4	2	0	0	2	0	0	2	2
$\chi^T(R') = f_R =$	3	1	1	-1	-1	-1	-1	-1	-3	-1	-1	1	1	1	1	1
$\chi^{(\omega)}(R') = f_R^{(C)} =$	3	1	1	-1	-1	-1	-1	-1	3	1	1	-1	-1	-1	-1	-1
$f_R^{\prime(1a)} = f_R^{\prime(B)} =$	2			0			-2	0	2			0			0	-2
$f_R^{\prime(1b)} = f_R^{\prime(B)} =$	2			0			0	-2	2			0			-2	0
$f_R^{\prime(2)} = f_R^{\prime(A)} =$	0	0	0	0	0	0	0	0	0	0	0	0	0	0	0	0
$\chi^{(\omega)}(R') = f_R'' =$	6	0	0	2	2	2	2	2	6	0	0	2	2	2	2	2
$\chi(R') = N_A^{PC}(R')f_R =$	24	2	2	-4	-2	-2	-6	-6	-6	0	0	6	0	0	4	4
$\chi^{TV}(R') = (N_M^{PC}(R')-1)f_R =$	9	1	1	-3	-1	-1	-3	-3	-3	1	1	1	-1	-1	1	1
$\chi^{RV}(R') = \sum_s N_{M_s}^{PC}(R')f_R^{\prime(s)} =$	4	0	0	0	0	0	-2	-2	4	0	0	0	0	0	-2	-2
$\chi^V(R') = \chi(R')-\chi^T(R')-\chi^{TV}(R')-\chi^{RV}(R') =$	8	0	0	0	0	0	0	0	-4	0	0	4	0	0	4	4
$\chi^{A_{1g}}(R') =$	1	1	1	1	1	1	1	1	1	1	1	1	1	1	1	1
$\chi^{A_{2g}}(R') =$	1	1	1	1	-1	-1	-1	-1	1	1	1	1	-1	-1	-1	-1
$\chi^{B_{1g}}(R') =$	1	-1	-1	1	1	1	-1	-1	1	-1	-1	1	1	1	-1	-1
$\chi^{B_{2g}}(R') =$	1	-1	-1	1	-1	-1	1	1	1	-1	-1	1	-1	-1	1	1
$\chi^{E_g}(R') =$	2	0	0	-2	0	0	0	0	2	0	0	-2	0	0	0	0
$\chi^{A_{1u}}(R') =$	1	1	1	1	1	1	1	1	-1	-1	-1	-1	-1	-1	-1	-1
$\chi^{A_{2u}}(R') =$	1	1	1	1	-1	-1	-1	-1	-1	-1	-1	-1	1	1	1	1
$\chi^{B_{1u}}(R') =$	1	-1	-1	1	1	1	-1	-1	-1	1	1	-1	-1	-1	1	1
$\chi^{B_{2u}}(R') =$	1	-1	-1	1	-1	-1	1	1	-1	1	1	-1	1	1	-1	-1
$\chi^{E_u}(R') =$	2	0	0	-2	0	0	0	0	-2	0	0	2	0	0	0	0

The rows $\chi^{A_{1g}}(R')$ through $\chi^{E_u}(R')$ are collected under $\chi^{(p)}(R') =$.

---

[10] The places left blank in the upper part of Table 18 refer to operations R which are not symmetry operations for the respective molecule. The corresponding values are not needed.

Table 19.

p	n_p	$n_p^T=n_p^\mu$	n_p^{TV}	n_p^{RV}	n_p^V	$n_p^{[\alpha]}$	$n_p^{\{\alpha\}}$	d_p	$d_p n_p$	$d_p n_p^T$	$d_p n_p^{TV}$	$d_p n_p^{RV}$	$d_p n_p^V$	$d_p n_p^{[\alpha]}$	$d_p r_p^{[\alpha]}$
A_{1g}	1	0	0	0	1	2	0	1	1	0	0	0	1	2	0
A_{2g}	2	0	1	1	0	0	1	1	2	0	1	1	0	0	1
B_{1g}	1	0	0	1	0	1	0	1	1	0	0	1	0	1	0
B_{2g}	1	0	0	0	1	1	0	1	1	0	0	0	1	1	0
E_g	2	0	1	1	0	1	1	2	4	0	2	2	0	2	2
A_{1u}	0	0	0	0	0	0	0	1	0	0	0	0	0	0	0
A_{2u}	3	1	1	0	1	0	0	1	3	1	1	0	1	0	0
B_{1u}	2	0	1	0	1	0	0	1	2	0	1	0	1	0	0
B_{2u}	0	0	0	0	0	0	0	1	0	0	0	0	0	0	0
E_u	5	1	2	0	2	0	0	2	10	2	4	0	4	0	0
Total								24	3	9	4	8	6	3		

We turn again to the method of Adams and Newton. We give first in Table 20 an excerpt from the data for space group No. 140 in their "Table 2".

Table 20.

140 $\mathfrak{D}_{4h}$										Rotatory										
Wyckoff	A_{1g}	A_{2g}	B_{1g}	B_{2g}	E_g	A_{1u}	A_{2u}	B_{1u}	B_{2u}	E_u	A_{1g}	A_{2g}	B_{1g}	B_{2g}	E_g	A_{1u}	A_{2u}	B_{1u}	B_{2u}	E_u
2 a	0	1	0	0	1	0	1	0	0	1	0	1	0	0	1	0	1	0	0	1
2 d	0	0	0	0	0	0	1	1	0	2	0	1	1	0	2	0	0	0	0	0
4 h	1	1	1	1	1	0	1	1	0	2	0	1	1	0	2	1	1	1	1	1

In this case we in addition need their "Table 4" partly extracted in Table 21.

Table 21.

140 $\mathfrak{D}_{4h}$	z Rotation									
Wyckoff	A_{1g}	A_{2g}	B_{1g}	B_{2g}	E_g	A_{1u}	A_{2u}	B_{1u}	B_{2u}	E_u
2 a	0	1	0	0	0	0	1	0	0	0
2 d	0	1	1	0	0	0	0	0	0	0
4 h	0	1	1	0	0	0	0	0	0	1

For n_p and n_p^{TV} the calculation proceeds as usual (use Table 20, left part):

			A_{1g} A_{2g} B_{1g} B_{2g} E_g A_{1u} A_{2u} B_{1u} B_{2u} E_u

n_p

			A_{1g}	A_{2g}	B_{1g}	B_{2g}	E_g	A_{1u}	A_{2u}	B_{1u}	B_{2u}	E_u
2 K-atoms on site a give	2a		0	1	0	0	1	0	1	0	0	1
2 N-atoms on site d give	2d		0	0	0	0	0	0	1	1	0	2
4 N-atoms on site h give	4h		1	1	1	1	1	0	1	1	0	2
	total		1	2	1	1	2	0	3	2	0	5

			A_{1g}	A_{2g}	B_{1g}	B_{2g}	E_g	A_{1u}	A_{2u}	B_{1u}	B_{2u}	E_u
$n_p^{TV}+n_p^T$	2 molecules M_2 on site a give	2a	0	1	0	0	1	0	1	0	0	1
	2 molecules M_1 on site d give	2d	0	0	0	0	0	0	1	1	0	2
		total	0	1	0	0	1	0	2	1	0	3
n_p^T	(from character tables)		0	0	0	0	0	0	1	0	0	1
n_p^{TV}	$= (n_p^{TV}+n_p^T) - n_p^T$		0	1	0	0	1	0	1	1	0	2

For n_p^{RV} complications arise from the fact that we have 2 <u>linear</u> molecules M_1 in the PC lying along adjacent edges (say directions x, y) of a square normal to the C_4-axis (= direction z). These molecules M_1 are the only ones responsible for rotatory vibrations. One molecule M_1 can rotate about x and z and the other one about y and z only. Now x and y by tetragonal symmetry are equivalent. We therefore can say that both molecules M_1 on site d can rotate about x and z, but not y. Now the numbers (let us call them briefly n_1) in Table 20 "Rotatory" correspond to rotational freedom about $x, y,$ and z, whereas those (say n_2) in Table 21 "z Rotation" to rotational freedom about z only. The difference $n_1 - n_2$ then accounts for rotational freedom about x and y, and, by the equivalence of x and y, half the difference for y only. Therefore we can calculate n_p^{RV} by taking $n_1 - \frac{1}{2}(n_1-n_2) = \frac{1}{2}(n_1+n_2)$ from rows 2d:

		A_{1g}	A_{2g}	B_{1g}	B_{2g}	E_g	A_{1u}	A_{2u}	B_{1u}	B_{2u}	E_u
	n_1	0	1	1	0	2	0	0	0	0	0
	n_2	0	1	1	0	0	0	0	0	0	0
n_p^{RV}	$= \frac{1}{2}(n_1+n_2)$	0	1	1	0	1	0	0	0	0	0

$n_p^{[\alpha]}$ and $n_p^{\{\alpha\}}$ again must be taken from character tables.

8.5. Example 5: A Hypothetical Structure

In order to discuss a case of particular complexity in a final example, we consider a hypothetical crystal structure with a primitive cubic PC Γ_c containing 8 identical diatomic molecules M = XY in a symmetrical arrangement at equal distances from the center and directed along the 4 space diagonals (see Fig. 6).

$\mathcal{S} = O_h^1 = Pm3m$ (space group No. 221), $\mathcal{P} = O_h = m3m$ (non-abelian), $[\mathcal{P}] = 48$.

Table 22 shows the sites and Tables 23, 24 summarize the FGA. The additional column 12a in Table 22 contains the values $\frac{\lambda}{l} g_A^\Lambda = \frac{\lambda}{l} \sum_s g_{A_s}^\Lambda$ which are immediately inserted in Table 23. Having only one molecular species M the index s on M_s can be omitted. M being linear, $f_R'^{(s)} = f_R'^{(B)}$ must be used; note that the axis of M is parallel to the reflection planes contained in its site group. In Table 23 C_3 stands for any one of the eight 120° rotations and σ (not σ_0 !) for any one of the six reflections occurring in Table 22, column 9.

Table 22.

1	2	3	4	5	6	7	8	9	10	11	12	12a	13	14	15	16	17	18
Λ	$\mathcal{L}$	$[\mathcal{L}]$	λ	l	$\frac{\lambda}{l}$	$\mathcal{L}^{(\mu)}$		$R \in \mathcal{L}^{(\mu)}$	A_s	$g_{A_s}^\Lambda$	$\frac{\lambda}{l}g_{A_s}^\Lambda$	$\frac{\lambda}{l}g_A^\Lambda$	λg_A^Λ	M	g_M^Λ	$\frac{\lambda}{l}g_M^\Lambda$		λg_M^Λ
g	$C_{3v} \equiv 3m$	6	8	4	2	$C_{3v}^{(1)}$		$E\ C_3^{(1)} \bar{C}_3^{(1)} \bar{\sigma}^{(1)} \bar{\sigma}^{(2)} \bar{\sigma}^{(3)}$	X;Y	1;1	2;2	④	16	M	1	②		8
						$C_{3v}^{(2)}$		$E\ C_3^{(2)} \bar{C}_3^{(2)} \bar{\sigma}^{(1)} \sigma^{(2)} \sigma^{(3)}$			2;2	④				②		
						$C_{3v}^{(3)}$		$E\ C_3^{(3)} \bar{C}_3^{(3)} \sigma^{(1)} \bar{\sigma}^{(2)} \sigma^{(3)}$			2;2	④				②		
						$C_{3v}^{(4)}$		$E\ C_3^{(4)} \bar{C}_3^{(4)} \sigma^{(1)} \sigma^{(2)} \bar{\sigma}^{(3)}$			2;2	④				②		

Table 23.

$\mathscr{P}' = \mathcal{O}_h' = \mathcal{O}_h^1/\mathscr{T}$	$k_q R' =$	1E'	8C'_3	3C'_2	6C'_4	6(C'_2)'	1i'	8S'_6	3σ'_0	6S'_4	6σ'
	$N_A^{PC}(R') =$	16	4	0	0	0	0	0	0	0	8
	$N_M^{PC}(R') =$	8	2	0	0	0	0	0	0	0	4
	$\chi^T(R') = f_R =$	3	0	-1	1	-1	-3	0	1	-1	1
	$\chi^{(\infty)}(R') = f_R'^{(C)} =$	3	0	-1	1	-1	3	0	-1	1	-1
	$f_R' = f_R'^{(B)} =$	2	-1								0
	$\chi^{(\alpha)}(R') = f_R'' =$	6	0	2	0	2	6	0	2	0	2
	$\chi(R') = N_A^{PC}(R')f_R =$	48	0	0	0	0	0	0	0	0	8
	$\chi^{TV}(R') = (N_M^{PC}(R')-1)f_R =$	21	0	1	-1	1	3	0	-1	1	3
	$\chi^{RV}(R') = N_M^{PC}(R')f_R' =$	16	-2	0	0	0	0	0	0	0	0
	$\chi^V(R') = \chi(R')-\chi^T(R')-\chi^{TV}(R')-\chi^{RV}(R') =$	8	2	0	0	0	0	0	0	0	4
	$\chi^{A_{1g}}(R') =$	1	1	1	1	1	1	1	1	1	1
	$\chi^{A_{2g}}(R') =$	1	1	1	1	-1	1	1	1	-1	-1
	$\chi^{E_g}(R') =$	2	-1	2	0	0	2	-1	2	0	0
	$\chi^{F_{1g}}(R') =$	3	0	-1	1	-1	3	0	-1	1	-1
$\chi^{(p)}(R') =$	$\chi^{F_{2g}}(R') =$	3	0	-1	-1	1	3	0	-1	-1	1
	$\chi^{A_{1u}}(R') =$	1	1	1	1	1	-1	-1	-1	-1	-1
	$\chi^{A_{2u}}(R') =$	1	1	1	-1	-1	-1	-1	-1	1	1
	$\chi^{E_u}(R') =$	2	-1	2	0	0	-2	1	-2	0	0
	$\chi^{F_{1u}}(R') =$	3	0	-1	1	-1	-3	0	1	-1	1
	$\chi^{F_{2u}}(R') =$	3	0	-1	-1	1	-3	0	1	1	-1

Fig. 6.

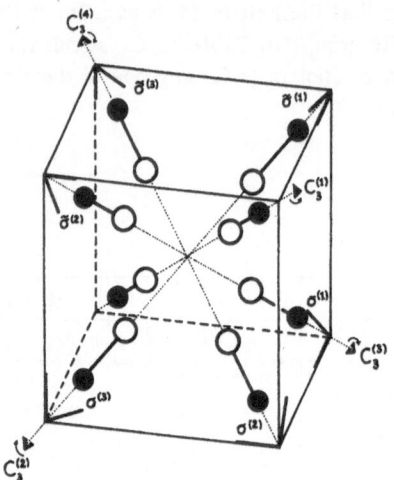

Table 24.

p	n_p	$n_p^T = n_p^M$	n_p^{TV}	n_p^{RV}	n_p^V	$n_p^{(or)}$	$n_p^{(\pi)}$	d_p	$d_p n_p$	$d_p n_p^T$	$d_p n_p^{TV}$	$d_p n_p^{RV}$	$d_p n_p^V$	$d_p n_p^{(or)}$	$d_p n_p^{(\pi)}$
A_{1g}	2	0	1	0	1	1	0	1	2	0	1	0	1	1	0
A_{2g}	0	0	0	0	0	0	0	1	0	0	0	0	0	0	0
E_g	2	0	1	1	0	1	0	2	4	0	2	2	0	2	0
F_{1g}	2	0	1	1	0	0	1	3	6	0	3	3	0	0	3
F_{2g}	4	0	2	1	1	1	0	3	12	0	6	3	3	3	0
A_{1u}	0	0	0	0	0	0	0	1	0	0	0	0	0	0	0
A_{2u}	2	0	1	0	1	0	0	1	2	0	1	0	1	0	0
E_u	2	0	1	1	0	0	0	2	4	0	2	2	0	0	0
F_{1u}	4	1	1	1	1	0	0	3	12	3	3	3	3	0	0
F_{2u}	2	0	1	1	0	0	0	3	6	0	3	3	0	0	0
							Total		48	3	21	16	8	6	3

Now we show how to solve this problem with the Tables of Adams and Newton. Table 25 is an excerpt for Wyckoff site g from their "Table 2" and Table 26 from their "Table 3". This latter "vector table" expressly has been calculated for representations based upon internal coordinates and advantageously can be used for the rapid derivation of the irrep numbers n_p^V of internal vibrations. In our case we have only stretching modes for the XY molecules. These modes can be represented by a set of 8 vectors directed from the cube center to the 8 corners. The site of the XY bond centers being g, the numbers n_p^V are immediately given by row 8g from "Table 3" reproduced in Table 26.

Table 25.

221 O_h											Rotatory
Wyckoff	A_{1g}	A_{2g}	E_g	F_{1g}	F_{2g}	A_{1u}	A_{2u}	E_u	F_{1u}	F_{2u}	(not needed)
8 g	1	0	1	1	2	0	1	1	2	1	

Table 26.

221	Vector									
Wyckoff	A_{1g}	A_{2g}	E_g	F_{1g}	F_{2g}	A_{1u}	A_{2u}	E_u	F_{1u}	F_{2u}
8 g	1	0	0	0	1	0	1	0	1	0

Thus the calculation proceeds as follows:

			A_{1g}	A_{2g}	E_g	F_{1g}	F_{2g}	A_{1u}	A_{2u}	E_u	F_{1u}	F_{2u}
n_p	8 X-atoms, site g (Table 25)	8g	1	0	1	1	2	0	1	1	2	1
	8 Y-atoms, site g (Table 25)	8g	1	0	1	1	2	0	1	1	2	1
		total	2	0	2	2	4	0	2	2	4	2
$n_p^{TV} + n_p^T$	8 mol. s M, site g (Table 25)	8g	1	0	1	1	2	0	1	1	2	1
n_p^T	(from character tables)		0	0	0	0	0	0	0	0	1	0
n_p^{TV}	$= (n_p^{TV} + n_p^T) - n_p^T$		1	0	1	1	2	0	1	1	1	1
n_p^V	8 bonds XY, site g (Table 26)	8g	1	0	0	0	1	0	1	0	1	0
n_p^{RV}	$= n_p - (n_p^{TV} + n_p^T) - n_p^V$		0	0	1	1	1	0	0	1	1	1

$n_p^{[\alpha]}$ and $n_p^{\{\alpha\}}$ again must be taken from character tables.

Acknowledgements. The author would like to express his gratitude to Dr. D. M. Adams, Leicester, for helpful discussions and a revision of the English text, and to Prof. J. Brandmüller, Dr. R. Claus, and Dr. H. W. Schroetter, Munich, for their active interest in this work.

References

1. S. Bhagavantam and T. Venkatarayudu: Proc. Ind. Acad. Sci 9A,
   224 (1939); — S. Bhagavantam: Proc. Ind. Acad. Sci. 13A, 543 (1941);
   — S. Bhagavantam and T. Venkatarayudu: Theory of Groups and its
   Applications to Physical Problems. New York: Academic Press, 1969.
2. J.-P. Mathieu: Spectres de Vibration et Symétrie des Molécules et des
   Cristaux. Paris: Hermann, 1945, Chap. XIII; — R. S. Halford: Journal
   of Chem. Physics 14, 8 (1946); — D. F. Hornig: Journal of Chem.
   Physics 16, 1063 (1948); — H. Winston and R. S. Halford: Journal of
   Chem. Physics 17, 607 (1949); — S. S. Mitra: Zeit. Krist. 116, 149
   (1961) and Sol. State Phys. 13, 1 (1962); — B. Schrader: Das Schwin-
   gungsspektrum von Molekülkristallen. Münster (University): Habilitationsschrift, 1968; — M. M. Sushchinsky: Spektry kombinatsionnovo
   rasseyaniya molekul i kristallov (Russ.). Moscow: Izd. Nauka, 1969,
   Chap. III, § 20; — H. Poulet et J.-P. Mathieu: Spectres de Vibration
   et Symétrie des Cristaux. New York: Gordon & Breach, 1970, Chap. V;
   — D. M. Adams and D. C. Newton: J. Chem. Soc. (A) 1970, 2822; —
   J. E. Bertie and J. W. Bell: Journal of Chem. Physics 54, 160 (1971); —
   J. E. Bertie and R. Kopelman: Journal of Chem. Physics 55, 3613 (1971);
   — W. G. Fateley, N. T. Devitt and F. F. Bentley: Appl. Spectrosc.
   25, 155 (1971); — L. L. Boyle: Acta Cryst. A27, 76 (1971), A28, 172
   (1972), Spectrochim. Acta 28A, 1347, 1355 (1972); — G. Turrell:
   IR and Raman Spectra of Crystals. New York: Academic Press, 1972,
   Chap. 4; — D. M. Adams: Coord. Chem. Rev. (in print).
3. F. Seitz: Ann. Math. 37, 17 (1936).
4. R. W. G. Wyckoff: Crystal Structures. New York: Wiley, 2nd edition
   1963sq, 6 Vols.
5. Landolt-Börnstein: Zahlenwerte und Funktionen (Numerical Data and
   Functional Relationships in Science and Technology). Berlin: Springer.
   Old Series, Vol. I, Part 4 (Kristalle), 1955; New Series Group III, Vol. 5,
   Parts a, b (Structure Data of Organic Crystals), 1971. (Other parts are
   to follow.)
6. N. F. M. Henry, K. Lonsdale: International Tables for X-Ray Crystallography. Birmingham: The Kynoch Press, Vol. I, 3rd edition 1969.
7. D. M. Adams and D. C. Newton: Tables for Factor Group and Point
   Group Analysis, 1970. (To be ordered from: Beckman-RIIC Ltd., Sun-
   ley House, 4 Bedford Park, Croydon CR9 3LG, England.)
8. E. B. Wilson jr., J. C. Decius, P. C. Cross: Molecular Vibrations. New
   York: McGraw-Hill, 1955.
9. J. L. de Boer, F. v. Bolhuis, R. Olthof-Hazekamp, A. Vos: Acta Cryst.
   21, 841 (1966).

10. A. Rosenzweig and B. Morosin: Acta Cryst. 20, 758 (1966).

11. R. Claus: Lichtstreuung an optischen Phononen und Polaritonen. Munich (University): Thesis, 1970, p. 76.

Dr. Josef Behringer
Sektion Physik der Universität München
Lehrstuhl Prof. Dr. J. Brandmüller
D-8000 München 40
Schellingstr. 4/IV
Federal Republic of Germany

SPRINGER TRACTS
IN MODERN PHYSICS

Ergebnisse der exakten Naturwissenschaften

Atomic Physics

Dettmann, K.: High Energy Treatment of Atomic Collisions (Vol. 58)

Donner, W., Süßmann, G.: Paramagnetische Felder am Kernort (Vol. 37)

Racah, G.: Group Theory and Spectroscopy (Vol. 37)

Seiwert, R.: Unelastische Stöße zwischen angeregten und unangeregten Atomen (Vol. 47)

Zu Putlitz, G.: Determination of Nuclear Moments with Optical Double Resonance (Vol. 37)

Elementary Particle Physics

Current Algebra

Furlan, G., Paver, N., Verzegnassi, C.: Low Energy Theorems and Photo- and Electroproduction Near Threshold by Current Algebra (Vol. 62)

Gatto, R. : Cabibbo Angle and $SU_2 \times SU_2$ Breaking (Vol. 53)

Genz, H : Local Properties of σ-Terms: A Review (Vol. 61)

Kleinert, H.: Baryon Current Solving SU (3) Charge-Current Algebra (Vol. 49)

Leutwyler, H.: Current Algebra and Lightlike Charges (Vol. 50)

Mendes, R. V., Ne'eman, Y.: Representations of the Local Current Algebra. A Constructional Approach (Vol. 60)

Müller, V. F.: Introduction to the Lagrangian Method (Vol. 50)

Pietschmann, H.: Introduction to the Method of Current Algebra (Vol. 50)

Pilkuhn, H.: Coupling Constants from PCAC (Vol. 55)

Pilkuhn, H.: S-Matrix Formulation of Current Algebra (Vol. 50)

Renner, B.: Current Algebra and Weak Interactions (Vol. 52)

Renner, B.: On the Problem of the Sigma Terms in Meson-Baryon Scattering. Comments on Recent Literature (Vol. 61)

Soloviev, L. D.: Symmetries and Current Algebras for Electromagnetic Interactions (Vol. 46)

Stech, B.: Nonleptonic Decays and Mass Differences of Hadrons (Vol. 50)

Stichel, P.: Current Algebra in the Framework of General Quantum Field Theory (Vol. 50)

Stichel, P.: Current Algebra and Renormalizable Field Theories (Vol. 50)

Stichel, P.: Introduction to Current Algebra (Vol. 50)

Verzegnassi, C.: Low Energy Photo and Electroproduction, Multipole Analysis by Current Algebra Commutators (Vol. 59)

Weinstein, M.: Chiral Symmetry. An Approach to the Study of the Strong Interactions (Vol. 60)

Electromagnetic Interactions

Deep Inelastic Lepton Scattering

Drees, J.: Deep Inelastic Electron-Nucleon Scattering (Vol. 60)

Landshoff, P. V.: Duality in Deep Inelastic Electroproduction (Vol. 62)

Llewellyn Smith, C. H.: Parton Models of Inelastic Lepton Scattering (Vol. 62)

Rittenberg, V.: Scaling in Deep Inelastic Scattering with Fixed Final States (Vol. 62)

Rubinstein, H. R.: Duality for Real and Virtual Photons (Vol. 62)

Rühl, W.: Application of Harmonic Analysis to Inelastic Electron-Proton Scattering (Vol. 57)

Experimental Techniques

Panofsky, W. K. H.: Experimental Techniques (Vol. 39)

Quantum Statistics

Graham, R.: Statistical Theory of Instabilities in Stationary Nonequilibrium Systems with Applications to Lasers and Nonlinear Optics (Vol. 66)

Haake, F.: Statistical Treatment of Open Systems by Generalized Master Equations (Vol. 66)

Semiconductors

Feitknecht, J.: Silicon Carbide as a Semiconductor (Vol. 58)

Grosse, P.: Die Festkörpereigenschaften von Tellur (Vol. 48)

Schnakenberg, J.: Electron-Phonon Interaction and Boltzmann Equation in Narrow Band Semiconductors (Vol. 51)

Superconductivity

Lüders, G., Usadel, K.-D.: The Method of the Correlation Function in Superconductivity Theory (Vol. 56)

X-Ray, Neutron-, Electron-Scattering

Steeb, S.: Evaluation of Atomic Distribution in Liquid Metals and Alloys by Means of X-Ray, Neutron and Electron Diffraction (Vol. 47)

Springer, T.: Quasi-Elastic Scattering of Neutrons for the Investigation of Diffusive Motions in Solids and Liquids (Vol. 64)

To Appear in Forthcoming Volumes:

Überall, H.: Study of Nuclear Structure by Muon Capture

Levinger, J. S.: Two-Nucleon and Three-Nucleon Systems

Brandmüller, J., Claus, R.: Light Scattering on Optical Phonons and Polaritons

Langbein, D.: Theory of van der Waals Attraction